The Incas' Sky

Émile Biémont

The Incas' Sky

From Myths to History and Astronomy

Émile Biémont
Xhendelesse, Belgium

ISBN 978-3-031-58417-6   ISBN 978-3-031-58418-3   (eBook)
https://doi.org/10.1007/978-3-031-58418-3

© The Editor(s) (if applicable) and The Author(s), under exclusive license to Springer Nature Switzerland AG 2024

This work is subject to copyright. All rights are solely and exclusively licensed by the Publisher, whether the whole or part of the material is concerned, specifically the rights of translation, reprinting, reuse of illustrations, recitation, broadcasting, reproduction on microfilms or in any other physical way, and transmission or information storage and retrieval, electronic adaptation, computer software, or by similar or dissimilar methodology now known or hereafter developed.
The use of general descriptive names, registered names, trademarks, service marks, etc. in this publication does not imply, even in the absence of a specific statement, that such names are exempt from the relevant protective laws and regulations and therefore free for general use.
The publisher, the authors and the editors are safe to assume that the advice and information in this book are believed to be true and accurate at the date of publication. Neither the publisher nor the authors or the editors give a warranty, expressed or implied, with respect to the material contained herein or for any errors or omissions that may have been made. The publisher remains neutral with regard to jurisdictional claims in published maps and institutional affiliations.

This Springer imprint is published by the registered company Springer Nature Switzerland AG
The registered company address is: Gewerbestrasse 11, 6330 Cham, Switzerland

If disposing of this product, please recycle the paper.

Preface

Who would not be seduced and intrigued by the marvelous Inca civilization that developed in the Cuzco basin in the heart of present-day Peru? The Inca empire was one of the largest and most impressive in pre-Columbian America, stretching from Colombia to Argentina and Chile, and covering most of present-day Ecuador, Peru, and western Bolivia. And there can be little doubt that it would have survived for a very long time had it not been destroyed by the Spanish conquistadors under the orders of Francisco Pizzaro.

The Inca people developed their extraordinary civilization, reaching such a high level of achievement, without ever having known the wheel or writing. For some of us, this stirs an irresistible urge to find out more about their culture. A few years ago, during a memorable trip to Peru, I was lucky enough to discover some of the most grandiose sites in this beautiful country and see for myself some of the apparently eternal traces of this astonishing empire. These ruins could hardly fail to impress, not only the anonymous visitor who happens upon these places, but even the experienced archaeologist.

Who could contemplate the grandiose and mystical site of Machu Picchu and remain unmoved? It seems that few can. Even the gods would be tempted to reside there. Anyone lucky enough to discover the marvelous archaeological sites of the pre-Columbian cultures that survive today will be struck by the grandeur and nobility of the Inca civilization, as revealed by several written testimonies, many in the form of Spanish chronicles. But it is important to understand that this impressive empire, which developed in territories of striking contrasts that some would describe as "cursed by the gods", was the culmination of the cultural contributions of a dozen astonishing and complex pre-Columbian civilizations that succeeded one another over the centuries.

This book first takes the reader on a guided tour of the climatic and geographical diversity of the regions where the Inca civilization took root. Then, through the testimonies of travelers, explorers, and archaeologists, and the writings of Spanish and Peruvian chroniclers, it presents the different cultures that preceded the Inca civilization, being careful to distinguish between its historical and mythical origins. Finally, we immerse ourselves in the spirituality and cosmology of the Inca people, built around their detailed and systematic observation of the sky, a common thread in the first few chapters.

The later chapters invite the reader to explore further the spirituality of the Inca civilization and understand how it relates to their observation of the sky and the practice of agriculture, so crucial to the survival of these populations. The history of the Inca people is characterized by an astonishing and omnipresent spirituality that cannot be ignored. They attributed a metaphysical power to many objects and places they considered sacred. These mystical places were deeply respected, particularly in the region around the capital Cuzco, where they were part of a system of seq'es radiating out from the Temple of the Sun.

Observation of the Sun was of great importance in Inca culture, particularly in determining the calendar and the agrarian cycles associated with the tropical year. As a consequence, observation of the sky plays a central role in this book. However, the reader need not be an expert in astronomy to grasp its significance. The peoples of the Inca kingdom observed the apparent motions of the Sun, the Moon, and the brightest planets. They were also familiar with certain particularly prominent constellations in the southern hemisphere, including bright star constellations and asterisms, interpreted in the Western manner, and 'dark' constellations with zoomorphic shapes standing out against the bright background of the Milky Way, itself likened to a sacred river.

Progressing through the book, the reader will discover the long and eventful history of Peru's different cultures and the civilizations that have succeeded one another in this often inhospitable land. I hope he will enjoy delving into the fascinating story of the Andean peoples, and the Inca world in particular. Finally, I hope the many photographs will bring to life the profound and ceaseless interaction between the heavens, the world of the gods, and the world of humans.

Note The transcription of Quechua terms is not always obvious and the spelling adopted can vary according to the authors. In this work, words of Spanish origin have been adopted. The names of sites and different cultures are usually written in Spanish. For common names, anthroponyms, and ethnonyms, the Quechua spelling has generally been preferred.

Spellings have been taken from the following two dictionaries:

- Gonzales O., Janney C.M., Thompson E.F., *Quechua–Spanish–English Dictionary*, Hippochrene Trilingual Reference, New York (2018),
- *Diccionario Quechua-Español-Quechua*, Academia Mayor de la Lengua Quechua, Gobierno Regional Cusco, Segunda edición, Cuzco (2005).

Xhendelesse, Belgium
February 2024

Émile Biémont
Honorary Director of the Belgian
FNRS, Member of the Belgian
Academy of Sciences

Contents

1	**Introduction**	1
2	**Geographical Contrasts and Climate Diversity of the Andean Regions**	5
	2.1 Landscape and Climate Diversity	5
	2.2 An Extreme Climate Episode: *El Niño*	9
3	**Sources Relating to Peruvian Civilizations**	13
	3.1 Famous Travelers, Explorers, and Archaeologists	13
	3.2 Exceptional Discoveries	18
	3.3 Spanish and Peruvian Chroniclers	20
	3.4 The *Khipukamayoqs* or 'Khipu Masters'	28
4	**History of Andean Civilizations**	33
	4.1 Chronology and Emergence of the Different Civilizations	33
	4.2 Chavín de Huántar and the Lanzón	36
	4.3 Lambayeque and Cupisnique Culture	37
	4.4 Paracas Culture and the *Fardos*	39
	4.5 Nazca Culture and Mummified Heads	40
	4.6 The Mystery of the Nazca Geoglyphs	43
	4.7 The Mochica Theocracy	49
	4.8 The Reign of the Mountain States: Tiwanaku and Wari	53
	4.9 Sicán and Chimú Civilizations	56
	4.10 Chancay and Chincha Cultures	61

5	**Birth, Evolution, and Splendor of the Inca Kingdom**		65
	5.1	The Inca Origin Myth	65
	5.2	The True History of the Inca People	70
	5.3	Pizarro and the Spanish Conquest	74
	5.4	The Last Gasps of the Empire	75
6	**Cosmogony at Tawantinsuyu and the Inca Pantheon**		77
	6.1	*Tawantinsuyu* or the World of the Four Districts	77
	6.2	Religious Practice and Beliefs of the Incas	81
	6.3	Inca Cosmogony	84
	6.4	The Complexity of the Inca Pantheon	85
7	**Proximity of the Gods and Sacred Places**		93
	7.1	Cuzco, the Flagship City of *Tawantinsuyu*	93
	7.2	On the Hill of *Saqsaywaman*	97
	7.3	Strange Rocks	99
	7.4	The Sacred Valley of the Incas	100
	7.5	Places of Pilgrimage	106
	7.6	Machu Picchu: At the Heart of the Sacred and Observation of the Sky	113
	7.7	Lake Titicaca and Surrounding Areas	125
8	**Measurement of Time, *Seq'es*, and Associated Rites**		129
	8.1	Synodic Period and Tropical Year	129
	8.2	The Motion of the Sun and the Rhythm of the Seasons	130
	8.3	The Inca Calendar	131
	8.4	The *Seq'es* of Cuzco	133
9	**Worship of the Sun**		143
	9.1	Introduction	143
	9.2	An Omnipresent Belief: The Cult of the Sun	143
	9.3	A Very Ancient Solstice Alignment	145
	9.4	A Sacred Territory	146
	9.5	Rocks and Astronomical Alignments	147
	9.6	Festivals and the Rhythms of Time	149
	9.7	Agriculture and Its Relationship with Observations of the Moon and Sun	150
	9.8	Horizon Astronomy and Archaeoastronomy	151
	9.9	Astronomical Observations at the *Qorikancha*	154
	9.10	Astronomical Observation Instruments and Sites	155

| | | Contents | xi |

10 Astronomy in the Andes — 159
   10.1 Hesiod and the Pleiades — 159
   10.2 The Sources of Inca Astronomy — 160
   10.3 Agrarian Cycles and Astronomy — 161
   10.4 The Role of the *Sukankas* — 163
   10.5 Gnomons and Observation of the Equinoxes — 164
   10.6 *Mayu*, the Celestial River — 168
   10.7 Archaeoastronomy and Ethnoastronomy — 169
   10.8 Uranometria — 170
   10.9 The Southern Cross — 171
   10.10 The Pleiades — 173
   10.11 The Constellation of Orion — 175
   10.12 From Urkuchillay to Pihca Conqui — 176
   10.13 Dark Constellations or *Yana Phuyu* — 178
   10.14 An Inca Zodiac? — 183
   10.15 *Ch'aska* and the Other Planets — 184
   10.16 The Andean Pilgrimage of *Coyllor Riti* — 185
   10.17 Andean Cosmology and the *Qorikancha* — 186
   10.18 When Wild Animals Attack the Moon — 188
   10.19 Comets and the Sadness of the Inca — 190

11 Conclusion — 191

Appendix A: Glossary of Spanish (S), Quechua (Q), and Aymara (A) — 197

Appendix B: Glossary of Astronomical Terms — 205

Appendix C: Brightest Stars in the Sky — 211

Appendix D: List of Constellations — 213

References — 217

Subject Index — 227

Index of Proper Names, Rulers, and Deities — 233

1

Introduction

From the Pacific Ocean to the Amazon forest, Peru's relief creates three contrasting geographical zones: the *costa*[1] (the coastal plains), the *sierra* (the Andean mountains), and the *selva* (the Amazon forest). A major corollary is the diversification of ecosystems. With such marked geographical transitions, the local populations had to be well adapted to the landscape. The intensive exploitation of harsh natural environments, essential for their survival, has favored a constant exchange of goods and ideas between the different peoples and a shared cultural tradition, despite local variants associated with different places and times. Andean history oscillates between tendencies towards standardization specific to vast 'temporal horizons,' with the dominance of clearly asserted cultures such as those of the Chavín, the Wari, or the Incas, and 'intermediate periods,' where regional specificities are manifested in particular through artistic creation.

The basic structure of Peruvian society over the centuries is the Andean clan or *ayllu*, which depends on mutual assistance in goods and services with the redistribution of available resources by the local chief, the *kuraka*. The *ayllu* regulates the use of land, organizing works of general interest and managing the problems associated with agricultural practices and generated by the production of goods. The Andean clan was the basic social structure of Inca society on which the power of the rulers was built.

The Inca people successfully integrated the cultural contributions of the societies that went before them to develop their domination over a huge territory and build up a highly effective political and social organization. Indeed,

---

[1] The words defined in Appendix A will be printed in italics in the text.

it worked so well because it was built on the contributions of previous models that were deeply anchored in the cosmological vision of the local populations. It took the brutal intervention of the Spanish conquistadors, bringing in their wake the inquisitors and the 'extirpators of idolatry,' acting in the name of the Church of Rome, to wipe out this beautiful civilization, which otherwise would probably have survived a great deal longer (Bernand, 2010).

Artistic production shows that Andean art was subject to many forms of political and religious constraints. This can be seen in the powerful functional architecture specific to temples, fortresses, and certain districts of cities, but also in the many and varied ceramics made for everyday or ceremonial use, monochrome or shimmering with color, radiant with the artisan's know-how. Not to mention the production of magnificent and expressive camelid wool textiles, which could be offerings dedicated to the divinities or the conspicuous sign of the high social status of their owners.

If we judge the Incas from what remains of their civilization today, including archaeological sites, written testimonies essentially in the form of Spanish chronicles, and an omnipresent iconography in the form of jewelry and ritual or everyday ceramics, what we find is that these people developed an extraordinary culture without inventing the wheel or writing. By making good use of human energy and intelligently implementing an extremely hierarchical administration, the Incas managed to develop in just a few decades a remarkably united society. Indeed, it was undoubtedly one of the most complex and admirable organizational structures the world has ever known.

Observation of the sky was of considerable importance for the Andean peoples, including the Incas and all the other civilizations that preceded them in this region. These populations observed the apparent motions of the Sun, the Moon, and the brightest planets. They were also familiar with the most prominent constellations in the southern hemisphere, whether constellations of bright stars forming asterisms or 'dark' constellations with zoomorphic shapes appearing in clear contrast to the diffuse glow of the Milky Way. This is attested by the numerous astronomical observations recorded in the writings of Spanish authors, but also in the abundant inscriptions found on steles and other monuments of a religious or secular nature at archaeological sites.

Studies of the arrangement and orientation of monuments in ceremonial centers as well as certain architectural alignments testify to the fact that these result from astronomical considerations, in particular from observation of the Sun (solstices and equinoxes). Their awareness of the heliacal rising and setting of certain stars, such as the Pleiades, suggests that all these peoples, and in particular the Incas, had a keen interest in observing the sky at certain significant times of the astronomical year. These observations and the development of calendar-related devices were justified by the need to plan agricultural tasks as the seasons went by, but they were also important for religious and ceremonial activities marking the significant moments of the tropical year.

The efforts made to obtain a detailed knowledge of certain astronomical phenomena are easy to understand when we consider the relationship between these phenomena and the environment in which these people lived, not to mention the cultural context. Consider, for instance, the survival problems caused by seasonal climatic changes linked to the *El Niño* phenomenon, a warm seasonal current off the coast of Peru and Ecuador that marks the end of the fishing season, and other climatic phenomena like drought, hurricanes, and storms. Astronomy provided a link between religious ideas and daily or seasonal agricultural practices, so it played an important role in Andean societies, where politics was also permeated by a lively cosmogony.

2
Geographical Contrasts and Climate Diversity of the Andean Regions

2.1 Landscape and Climate Diversity

We generally include in the region called the 'Central Andes' an area delimited by the Pacific Ocean on one side and by the Amazon rainforest on the other. This region includes the far north of Chile, the highlands of Bolivia, and present-day Peru, which encompasses the coastal zone, the central mountains, and the Amazonian piedmont. Its northern limit is located not far from the border between Peru and Ecuador. On the southern borders, we find the high plateaus which extend the region of Lake Titicaca towards the south, where we encounter almost uninhabited areas such as the Atacama Desert.

Except for the southern highlands, the Central Andes is a large but highly fragmented region featuring many narrow coastal valleys with steep slopes, separated from each other by desert areas (Fig. 2.1). The traveler going from the Pacific Ocean towards the Amazon forest will encounter, over a distance of approximately 200 km, a wide variety of different ecosystems, including essentially three types of environment arranged according to a vertical structure. The hot lands extend on both sides of the Andes up to an altitude of around 2000 m, although this varies depending on the location. Higher up, there are mountain valleys with crops present up to around 3500 m. Higher still, and up to around 5000 m, the landscape is made up of the cold areas of the *puna*, which finally give way to rock formations and glaciers with the possible presence of moss and lichens. Some mountain peaks exceed 6000 m in altitude.

The coastal strip, with the Atacama Desert, forms a desert region where it hardly ever rains, except for a light drizzle (*garúa*) in winter. It also has many cliffs covered in cacti. In this region, we sometimes find deep valleys with fertile

Fig. 2.1 Desert landscape along the road from Lima to Nazca. Author's photograph

and largely irrigated areas as well as oases where cotton, corn, and squash are cultivated (Figs. 2.2 and 2.3). Above 300 m, the temperature is favorable for growing avocados, guavas, chili peppers, lucumas, and coca.

Virgin forest (or *selva*) grows in the eastern foothills of the Andes. It covers the Amazonian lowland and corresponds to relatively sparsely populated areas. The inhabitants currently cultivate peppers, cotton, coca, peanuts, cocoa, cassava, and avocado. There are also precious woods, resins, and pharmacological plants. The *selva* has a humid, tropical climate with up to 200 days of rain per year and temperatures easily reaching 30 °C.

In the mountains (or *sierra*), farmers have developed intensive agriculture by setting up numerous cultivable terraces on the slopes at the temperate level. These areas are favorable for agricultural production of quinoa, amaranth, barley, potatoes, and oca, crops grown up to 4000 m above sea level. Farmers have long used irrigation to compensate for variations in rainfall (Fig. 2.4).

The Andes Mountain Range (*Cordillera de los Andes*) is the north–south oriented mountain range that extends along the western coast of South America. It is approximately 7000 km long and varies between 200 and 800 km wide. The cordillera peaks at 6962 m and its highest point in Peru is the Huascarán, at 6768 m. Many peaks are snow-capped. The Andes are part of the Pacific Ring of Fire, an area of intense volcanic activity (Fig. 2.5).

The *puna* is a grassland region specific to high altitude, stretching south of the Central Andes. It borders the arid and often desert coast of the Pacific. The *puna* is a particularly good area for breeding llamas and alpacas. There

Fig. 2.2 An example of a fertile valley where intensive agriculture is practised near desert areas. This photo was taken along the road from Arequipa to Nazca. Author's photograph

Fig. 2.3 Oasis of Huacachina (province of Ica). Author's photograph

Fig. 2.4 High mountain landscape with deep valleys and snow-covered peaks. Photo taken near Colca Canyon. Author's photograph

is also an abundance of deer, rodents, and camelids such as the vicuña, but also predators like the puma (Fig. 2.6). In places where there is no night frost, quinoa and a wide variety of potatoes are grown, but also more specific crops such as oca, ulluco, mashua, and maca.

In the high plateau of the Central Andes is Lake Titicaca, whose banks are very fertile. In these regions, the flooded field technique was once used for agriculture, a fertilization method which at one time ensured the development of Tiwanaku (Mathé 1996; Cavatrunci et al. 2005).

Peru has two seasons: a dry season and a wet season. In the Amazon plain and on the eastern slopes of the Andes, the heaviest precipitation occurs from January to April, while the dry season extends from May to November. Summer is generally a little cooler than winter. In the mountains, the months of January to April are often hotter and more humid than the months of May to September, which are usually dry and cool. In the desert region, summer occurs from December to March. At this time, the air is hot and humid, but from May to November, a mist that the Peruvians call *garúa* forms along the coast.

The Northern Andes, in the Quito region, have a different appearance from the Central Andes. At high altitudes (above 3500 m), wet meadows are favorable for livestock farming, particularly alpaca and llama, and there is a large population of farmers in the mountain valleys (Rachowiecki & Beech 2006).

Fig. 2.5 The Peruvian Andes cordillera seen from an airplane. Author's photograph

2.2 An Extreme Climate Episode: *El Niño*

The climate in the coastal regions of Peru is influenced by sea currents. Normally, the coasts of Chile, Peru, and Ecuador are bathed in the cold current known as the Humboldt, which heads towards the north, and they are swept by winds which blow from the south-east towards the north-west. These winds chase away warm water from the ocean surface, which generates an upwelling of colder water from a depth of between 100 and 200 m. These waters are rich in nutrients and cause the development of plankton, which attracts birds and fish, a favorable context for the development of fishing activities.

Every year, around Christmas and until April, a coastal current from the open sea is set in motion towards the south. This is a warm current called *El Niño* ('little boy,' in Spanish), which appears off the coast of Peru and Ecuador

Fig. 2.6 The *puna* is the ideal area for breeding llamas and alpacas. Author's photograph

and which is characterized by abnormally high temperatures (Changnon & Bell 2000). Its appearance moves the areas of precipitation eastward into the Pacific Ocean, and prevents the upwelling of cold water along the coast of South America. This results in a depletion of nutrients in the water, with a reduction in wildlife and significant collateral damage for the fishing industry. At the same time, the coastal regions of northern Peru and Ecuador, which usually receive little rain, experience extremely heavy rainfall.

The name *El Niño* refers to the Child Jesus and is attributed by South American fishermen because it occurs shortly after Christmas. Originally, this term referred to an annual climate episode before being adopted for 'extreme' situations with a very warm current which manifests itself further south up to the coast of Chile. Very marked events of this type, with temperature anomalies of up to 4 or 5 °C, took place in particular in 1982–83, 1997–98, and 2014–16. *El Niño* generally lasts about eighteen months (Glantz 1996; Caviedes 2002).

La Niña (the 'little girl,' in Spanish) refers to an opposite phenomenon to *El Niño*, but constitutes an important episode of climate variability. *La Niña* episodes occur every 4 to 5 years, generally last 1 to 2 years, and do not appear to be directly correlated with *El Niño*. This climate disturbance, which finds its origin in a thermal anomaly in the equatorial surface waters of the central Pacific Ocean, is characterized by an abnormally low temperature of these waters.

3

Sources Relating to Peruvian Civilizations

3.1 Famous Travelers, Explorers, and Archaeologists

Important information about the Andean peoples comes from travelers who criss-crossed South America at different times and who have supplied us with first-rate documentation relating to the customs and traditions of the original inhabitants of these regions.

There are a great many ancient sites in Peru. Archaeological discoveries continued in this country throughout the twentieth century, revealing numerous traces of the first civilizations, and a number of museums have been built to house the finds. On the basis of the ruins that have been unearthed, different theories have been proposed or explored further, and this has contributed to a better knowledge of the history of the country and the various civilizations that developed on its territory. These contributions are the work, on the one hand, of foreign scientists who came to carry out excavations in Peru, but also of Peruvian archaeologists, stimulated by the numerous discoveries in their own country, sometimes spectacular and widely publicized.

In a sense, we may say that modern archaeology began in Peru when the German geologist and explorer William Reiss (1838–1908), along with his compatriot, the geologist and volcanologist Alphons Stübel (1835–1904), stumbled across a huge necropolis on the coast bordering Ancón Bay. These ruins were discovered in 1875 during the construction of the Lima–Chancay railway. In fact, in 1868, Reiss embarked with Stübel on a trip to Hawaï but both stopped in Colombia, fascinated by the Andes. They then carried out a series of geological, ethnographic, and archaeological studies in Colombia, Ecuador, and

Peru. Back in Germany, the two scientists published in Berlin (Reiss & Stübel 1880–1887) *Das Totenfeld von Ancón in Peru*, a monumental text which was translated into English under the title *The Necropolis of Ancón in Peru*. Stübel went on to carry out further research on the Tiwanaku site in Bolivia.

Friedrich Maximilian Uhle, abbreviated Max Uhle (1859–1944), was a German archaeologist who is considered the father of Peruvian archaeology. From 1888, as an assistant at the Berlin Museum, he became interested in South American archaeological collections. From 1892, he studied certain archaeological sites in Bolivia and Peru and became interested in the collapse of the Inca empire. With funding from several American universities, he visited different places and took part in excavations at various sites such as Tiwanaku, Pachakamaq near Lima, Chancay, and Supe. He was also interested in Chimús and Moche ruins in the north of the country and made many original contributions on the archaeological and ethnological levels.

From 1905, he visited a site near Caral, and subsequently carried out further excavations in Chile and Ecuador, before returning to Germany in 1933. In 1935, he published a synthesis of his contributions under the title *Die alten Kultur von Peru*, a work devoted to pre-Inca civilizations. A Spanish version of this text was published in 1956 (Uhle 1956). He was the first to distinguish the Mochicas and the Chimús on the cultural level, and in 1913, wrote a text entitled *Die Ruinen von Moche* (Uhle 1913). Stübel also worked with Max Uhle at Tiwanaku and published with him an important work related to this site: *Die Ruinenstätte von Tiahuanaco im Hochlande des alten Peru* (Uhle & Stübel 1892). There is a work on Max Uhle and his contributions to Peruvian archaeology by J. H. Rowe (Rowe 1954).

John Howland Rowe (1918–2004) was an American archaeologist and anthropologist. His notoriety comes mainly from the studies he published relating to the Inca civilization. After studying at Brown University and Harvard University, he undertook excavations notably in southern Peru, then in Wari and Cuzco, under the auspices of the Peabody Museum of Archaeology and Ethnology of Harvard University. He developed departments of archaeology and anthropology in the universities where he stayed (Cuzco, Popayan in Colombia) then he taught at the University of California at Berkeley until the end of his professional life in 1988. His work on our understanding of Inca culture at the time of the Spanish conquest (Rowe 1946, 1957) constitutes a major contribution to this field.

Rowe proposed a chronology of pre-Hispanic Peru which still remains a reference for Peruvian archaeology today. The 'periodization' of the pre-Columbian era that we owe to him is based on changes in the style of ceramics and remains a major reference. To classify the different archaeological phases, he refers to 'cultural horizons' and 'intermediate periods.' The horizons designated periods during which a dominant culture developed and exerted its influence over the entire country while the intermediate periods were phases during which the regions regained a certain autonomy and their social and religious particularities. Established in 1958, this classification could not, however, take into account subsequent discoveries such as those made at the site of the Kotosh temple or on 'pre-ceramic' sites dating back more than 1000 years BCE.

Adolph Francis Alphonse Bandelier (1814–1940), an American archaeologist and anthropologist of Swiss origin, was a great specialist in pre-Columbian civilizations. He was particularly interested in Ecuador, Peru, and Bolivia, and collaborated with the Museum of Natural History in New York. In 1911, he was commissioned by the Carnegie Institute in Washington to study, particularly in Spain, the archives relating to Indian populations. Between 1892 and 1903, he carried out numerous excavation campaigns in the Peruvian altiplano and in Bolivia, particularly in the region of Lake Titicaca, and in 1919, he published the work entitled *The Islands of Titicaca and Koati* (Bandelier 1910). He studied in detail the citadel of Kuelap, the capital of the Chachapoya people. Built around the nineteenth century, this fortress was the largest construction built by these people. It is a spectacular structure covering an area of more than 7 ha and is protected by a precipice and narrow entrance corridors. The studies made by Bandelier, who is among the pioneers of pre-Columbian archaeology, are based not only on excavations but also on archival investigations, and give pride of place to ethnological and anthropological methods.

Hiram Bingham (1875–1956) was an American explorer and politician. Born in Honolulu, he graduated from Yale in 1898, the University of California in 1900, and Harvard in 1905. He taught history and politics at Harvard and then at Princeton. In 1909, he went to Peru and learned about Inca culture. It was in 1911 that he discovered the site of Machu Picchu while searching for the ruins of Vilcabamba, the last refuge of the Incas after the Spanish conquest. He discovered the main access routes to the site, uncovered numerous tombs, and exhumed a large number of objects of archaeological interest. His discovery, published by *National Geographic* magazine in 1913, was a resounding success. In 1948, he published *Lost City of the Incas, the Story of Machu Picchu and Its Builders* (Bingham 1948), which is still one of the major works on the Machu Picchu site. This text is also available in French (Bingham 1989, 2008).

It seems likely that H. Bingham only re-discovered the Machu Picchu site. According to a Peruvian Benedictine historian and explorer of the Cuzco region, Paolo Greer, the primacy of discovery should go to the German Augusto Berns, a mining prospector who worked in the Cuzco region in 1860–1870 and who owned land encompassing Machu Picchu. He even obtained authorization, in 1887, to exploit a *waka* (sacred place) located on this land.

Alfred Louis Kroeber (1876–1960) was an American anthropologist. He spent most of his life at the University of California at Berkeley and worked alongside the Native Americans of California for the recognition of their rights. He made notable contributions to archaeology and anthropology, but is best known for his work in ethnology. During the years 1925–1926, he carried out excavations in different regions of Peru, including the valleys of Lima, Cañete and Rio Chillon, but also in the regions of Nazca and Paracas and on the Chimús sites. In the 1930s, he provided the basis for a historical classification of Peruvian civilizations obtained from the study of ceramics preserved at the University of Berkeley. In this institution, in 1945, he founded the Institute of Andean Research which achieved great notoriety.

Junius Bird (1907–1982) was an American archaeologist who was appointed curator of the department of South American archaeology at the American Museum of Natural History in New York in 1934. He participated in expeditions to Chile during the period extending from 1934 to 1942 and then, in 1946–1947, carried out studies on Paijan man (the pre-ceramic period) in northern Peru. He also undertook research on the site of Huaca Prieta, where he discovered the oldest ceramics on the north coast. This brought to light the first developments in agriculture and activities related to the manufacture of very ancient textiles. These discoveries established the first carbon-14 datings and an absolute chronology for the formative period during which the first villages and political organizations developed, favoring the appearance of theocratic structures.

Among the more important contributions to the study of the Inca world, we can also cite those of William H. Prescott (1796–1859), an American historian specializing in the history of the Hispanic world, who published, in 1847, *History of the Conquest of Peru* (Prescott 1847, 2005). Another important work by this author was published under the title *Aztèques et Incas. Grandeur et décadence de deux empires fabuleux* (Aztecs and Incas. The Rise and Fall of Two Fabulous Empires) (Prescott 2007).

While the discoveries of foreign scientists stimulated national interest in the excavated sites, they also contributed to the emergence of a generation of Peruvian archaeologists including Julio César Tello Rojas, Jorge C. Muelle, and Rafael Larco Hoyle.

Julio César Tello Rojas (1880–1947) is considered one of the great pioneers of Peruvian archaeology and is undoubtedly the best-known and most respected Peruvian archaeologist. Until 1947, this doctor and anthropologist directed the National Museum of Archaeology, Anthropology, and History of Peru (*Museo Nacional de Arqueología, Antropología e Historia del Perú*). He was keen to advertise the ancestral greatness of the Peruvian people and the different cultures that succeeded one another on the territory of his country. He gave objective support to his theories by basing them on numerous excavations he carried out and discoveries he made in Chavín, Paracas, Wari, and in many other places in the Andes or along the Pacific coast (Burger 2009). He considered the culture of Chavín as the mother culture of all the civilizations of Peru and believed that the common core of his country's many cultures was the Amazonian regions. From there, people gradually spread into the Andean valleys in search of better living conditions. According to him, the cultures of the eastern Andes were the oldest, then those of the western Andes, and finally those of the coast developed from the previous ones, all of which were finally absorbed within the Inca empire. It was Julio César Tello Rojas who founded the Museum of Archaeology and Ethnology at the University of San Marcos.

Rafael Larco Hoyle (1901–1966) was a Peruvian archaeologist born in Trujillo. Graduating as an agricultural engineer from Cornell University in the United States, he returned to Peru from 1923 to 1956 to take charge of the family hacienda. It was the purchase by his father, Rafael Larco Herrera, of a collection of vases and ancient ceramics that sparked Larco Hoyle's enthusiasm for archaeology. This collection would later become the embryo of the future *Museo Arqueológico Rafael Larco Herrera*, which was set up in Lima in 1958 (Fig. 3.1).

Rafael Larco Hoyle carried out numerous excavations and made many discoveries, using which he was able to build up an extraordinary collection of more than 45,000 pieces, first in the Chiclin hacienda in Trujillo, then in the Lima museum. This collection contains works from the Chimú civilizations, and the Lambayeque, Inca, Mochica cultures, and others. Among his major contributions, Larco Hoyle established in 1944 a coherent chronology of northern Peru and then of the whole of Peru divided into seven eras. He also discovered the Viru, Salinar, and Cupisnique cultures, and wrote a monumental work on the culture of the Mochicas, the most diverse aspects of which he studied in detail. He opposed the theories of Tello Rojas, who made the Chavín civilization the matrix of all pre-Columbian civilizations (Larco Hoyle 1938, 1940).

Jorge C. Muelle (1903–1974) was a Peruvian anthropologist and archaeologist, and a disciple of Max Uhle. He studied at the National University of Trujillo and at the Major National University of San Marcos. He also studied at the universities of Yale and Berkeley. He was notably the director of the National Museum of Archaeology, Anthropology, and History of Peru. He is

Fig. 3.1 View of the Rafael Larco Herrera Museum in Lima, which was founded by the Peruvian archaeologist Rafael Larco Hoyle. Author's photograph

known as a specialist on the coastal civilizations of Peru, and carried out excavations in Paracas (1931), in the Lima region (1935), and at Pacaritambo in the vicinity of Cuzco.

3.2 Exceptional Discoveries

The end of the twentieth century and the beginning of the twenty-first century were marked by resounding discoveries which helped to publicize interest in Peruvian archaeology.

The archaeological site of Kotosh is located 1950 m above sea level in the Central Andes, 5 km from the town of Huánuco in the Huallaga Valley. The site was excavated from 1960 to 1966 by a team from the University of Tokyo led by Seichi Izumi (1915–1970). Japanese archaeologists found around ten superimposed constructions in a mound 13 m high and 100 m in diameter. The oldest building has thick walls with trapezoidal niches. It is known as the 'crossed hands' temple because a bas-relief in dried clay representing two crossed forearms, dating back to pre-ceramic times, was discovered there in 1963. In the White Temple, near the Temple of the Crossed Hands, clay offerings were also found. These two constructions are the oldest known ceremonial buildings in Peru and would correspond to a late pre-ceramic phase (approx. 2500–1800 BCE), characterized by the use of stone tools and stone figurines or clay.

La Huaca Rajada is a funerary complex of the Mochica culture located near the village of Sipán, 30 km from Chiclayodans, in the Lambayeque region of northern Peru. This place became famous thanks to the discovery of the tomb of a high-ranking Moche leader called 'Lord of Sipán' in 1987–1988 by Walter Alva (born in 1951) and his wife Susana Meneses de Alva (1948–2002), both specialists in Mochica culture. This discovery was especially important because it was one of the rare pre-Columbian sites that had not been visited by the *huaqueros*, or tomb robbers. The Lord of Sipán was buried in the company of two men, two women, and a dog. The tomb contained many precious objects and ornaments made of gold, silver, copper, and bronze. It gave archaeologists a better understanding of the ceremony of sacrifice of the sovereigns of Sipán, often illustrated on ceramics or wall paintings. The finds from this funerary complex are kept in the *Tumbas Reales de Sipán* museum in the town of Lambayeque.

The Caral civilization, also referred to as the Norte Chico civilization or the Caral-Supe civilization, is a pre-ceramic culture belonging to the pre-Columbian Late Archaic Period. It developed in the Norte Chico region, in the center of the northern coast of Peru, about 200 km from Lima, and flourished between the thirtieth and twenty-eighth centuries BCE. It is characterized by monumental architecture comprising raised platforms and hollow circular squares, but there were no artistic creations or ceramics. Although the presence of ancient sites in these regions had been known for a long time, we owe the first in-depth study of this civilization, at the end of the 1990s, to Peruvian archaeologists led by Ruth Shady Solis (born in 1946), thanks to excavations carried out at Caral. It was the work published in the early 2000s by this Peruvian personality, who was curator, then director of the National Museum of Archaeology, Anthropology, and History of Peru, which drew public attention to the importance of these sites. We are lost in conjectures about the reasons for the disappearance of this civilization between 1900 and 1800 BCE, which could be due to very violent earthquakes or extreme climatic episodes, possibly linked to *El Niño*.

An American explorer, archaeologist, and anthropologist who graduated from the University of Vienna (1974), Johan Reinhardt (born in 1943), devoted part of his life to studying the Nazca Lines, but also the ceremonial centers of Peru, including those of Chavín, Machu Picchu, and Tiwanaku. His numerous climbs in the Andean mountains allowed him to discover more than fifty Inca ritual sites located at high altitude. In 1995, he led the expedition which unearthed, in the ice of Nevado Ampato (6300 m above sea level), the mummy of a young girl sacrificed on the mountain. She was called Juanita, and nicknamed the Ice Maiden, and is currently kept in a museum in Arequipa. His expeditions to the Andes, during the period 1996–1999, led to the discovery of other mummies and sacrificial sites in the mountains. He published several papers reporting his discoveries (Reinhardt 1998, 2005, 2007).

The Peruvian archaeologist, anthropologist, and historian Frederico Kauffmann Doig (born in 1928) was notably director of the Lima Art Museum and the National Museum of Archaeology, Anthropology, and History of Peru. Author of numerous archaeological and historical works, he has made important contributions to the study of pre-Columbian civilizations, including the culture of Chavín and the culture of the Chachapoya people (see, for example, Kauffmann Doig 2005). He carried out research in particular in Arequipa, Ica (the painted temple of El Ingenio in Nazca), and Lima (the site of Ancón). He focused particularly on the Chachapoya culture. As testimony of this culture, in 1997, he brought to light the mummies of Leymebamba or the Lake of the Condors (referred to today as Laguna de las Momias). They were found in a vast necropolis built on cliffs and containing around 280 mummies preserved in burial chambers (*chullpas*). These mummies are now in a museum built for their conservation in Leymebamba. Around three thousand objects, including fabrics, necklaces, ceramics, and *khipus* from the Inca or pre-Inca period, were also discovered at this site.

3.3 Spanish and Peruvian Chroniclers

The past of the Andean peoples is known to us thanks to numerous historical or ethnological documents that have been published (García 2000). Many of them come from Spanish or South American chroniclers who lived in Peru at the time of the arrival of European colonizers or immediately after (de Castro Yupangui 2005). The first chroniclers were conquistadors or ecclesiastics, often Jesuits, who participated in the evangelization of the country. They were followed by descendants of local nobles, who wrote texts describing the history and customs of their people, but also sometimes, the living conditions of the indigenous peoples during the colonial era (Karsten 1993).

We give below a list of the main works or documents that have been published and a brief summary of the life and work of different chroniclers who have contributed to making us aware of the history and traditions of the various civilizations of Peru and Inca culture in particular. These authors are classified chronologically according to when they lived.

Juan Polo de Ondegardo y Zárate (ca. 1500–1575) arrived in Peru around 1546 and lived in this country for about thirty years until his death. After the promulgation of the laws of 1542 limiting the power of the *Encomenderos*[1], he participated in the war against the rebels and obtained different promotions after the fall of Gonzalo Pizarro in 1548. He was *corregidor* (justice officer) of Cuzco twice from 1558 to 1560 and from 1571 to 1572, while Francisco de Toledo was viceroy. He participated in many important events of the viceroyalty during the second half of the century. Polo de Ondegardo wrote a text entitled *Los errores y supersticiones de los Indios* [1585] (1916) and another which bears the title *Relación de los fundamentos acerca del notable daño que resulta de no guardar a los indios sus fueros* [1571] (1872). His texts are mainly reports to civil and religious authorities and reflect the author's detailed knowledge of the indigenous societies. They give us a better understanding of the religion and beliefs of the indigenous peoples. The writings of Polo de Ondegardo shed light on the relationships between local populations and the Spanish during the early stages of colonization.

Juan Diez de Betanzos (1510–1576) wrote the *Suma y narración de los Incas* [1551] (1987), which is an important source of information regarding the Inca civilization. His story is based on the testimonies of his wife Cuxirimay Oclllo (Doña Angelina), who had been the wife of the king *Atawallpa*, and on those of soldiers who took part in the battle of Cajamarca. The story reflects the indigenous point of view and provides information regarding the development of the Inca empire during the reign of *Tupaq Inca Yupanki*, the expansion due to *Wayna Qhapaq*, and the civil war between *Waskar* and *Atawallpa*. This story was written for the viceroy of Peru, Antonio de Mendoza. De Betanzos was one of the first Spanish settlers to settle in Cuzco. He knew Quechua well and had contacts with members of the old ruling class. His story is therefore of particular importance.

Francisco de Toledo (1515–1582) was the fifth viceroy of Peru, in office from 1569 to 1581. In 1570, he installed the tribunal of the Inquisition. He crushed the revolt of *Tupaq Amaru*, whom he had executed. He centralized the colonial administration and established the foundations of Peru's administrative system. In his letter to the King of Spain Philip II, he mentions the existence of a report on the *wakas* sent to Spain in 1572.

---

[1] The *encomienda* was, during the conquest of the New World, a system applied by the Spaniards for economic and evangelization purposes. It consisted of forcing the natives to work without pay in mines and fields, or for works of public interest.

Juan de Matienzo (1520–1579) was a distinguished jurist who spent part of his life in Peru. He wrote *Gobierno del Perú* [1567] (1967), a treatise which presents a reform of the Spanish political system in force in the Andes. He advocated the census and destruction of the *wakas*. His manuscript seems to indicate that he was not the author of the text, which was later copied by Cobo, but which could have been written by Polo de Ondegardo.

Pedro Cieza de León (1520–1554) was a Spanish chronicler and conquistador who visited Cuzco in 1550. In 1553 and after his return to Spain, he published in Seville the first part of *Crónica del Perú* [1553] (1984) (Fig. 3.2). He participated in several military campaigns led by Pedro de la Gasca against the rebellion of Gonzalo Pizarro. It was in 1548 that he settled in what is now the city of Lima and began his career as a chronicler of the New World, which

Fig. 3.2 First part of the *Crónica del Perú* by Pedro de Cieza de León, dated 1553. University of Pennsylvania Library, public domain

allowed him to travel through Peruvian territory for two years. He died shortly after the publication of the first volume of his *Crónicas*. The second part was published under the title *Segunda parte de la Crónica del Perú, que trata del señorío de los incas yupangueis y de sus grandes hechos y gobernación* (1871). The last part entitled *Tercer libro de las Guerras Civiles del Perú, el cual se llama la Guerra de Quito* appeared in Madrid in 1909.

Martin de Murúa (ca. 1525–ca. 1618) was a Mercedarian friar and chronicler of the Spanish conquest of the New World. He published a *Historia general del Perú*, which was written from 1580 [ca. 1615] (2008). It is undoubtedly the first illustrated history of Peru. This author stayed in different places including Cuzco, Arequipa, and around Lake Titicaca. He returned to Spain in 1615 after a long detour through Argentina. His book concerns pre-Columbian Peru, but also the beginnings of the colonial era. There are two versions of the *Historia general del Perú*.

Cristóbal de Molina (1529–1585) ('El cuzqueño') was a Spanish clergyman and chronicler who resided in Cuzco for a long time because he loved Andean culture. It was in 1556 that he settled in Peru and learned Quechua. He is known for having comforted the Inca *Tupaq Amaru* during his execution in the parade ground of Cuzco in 1572. He wrote the *Relación de las fábulas y ritos de los Incas* [1575] (1989). It is his main work and the only one preserved. In the first part of this work, we find a description of Incan myths and legends, particularly those relating to their origins, while the second part of the work describes Incan rites and beliefs.

Felipe Guamán Poma de Ayala (ca. 1530–ca. 1615) was a Peruvian chronicler of the time of the Spanish conquest. Despite his mixed name, he was indigenous and of a noble Inca family. Of Quechua identity, he also considered himself of Latin origin and Christian. He inherited the name of conquistador Luis Avalos de Ayala after fighting alongside him and saving his life. He visited many places in Peru including Cuzco and Lima. His work entitled *El Primer Nueva Crónica y Buen Gobierno* was completed around 1615 [1615] (1936, 1980). It is a sort of petition addressed to the King of Spain, richly illustrated, and whose drawings describe in great detail the living conditions of the indigenous peoples of Peru after the Spanish conquest. The language he uses is sometimes difficult to decipher. This work represents a very valuable source of information about the lives of the indigenous peoples during the time of the Inca empire. Guamán Poma de Ayala was able to gather important information from the archivists of the Inca empire (*the khipukamayoqs* or 'masters of the *khipu*') during his numerous travels in the Andes. This document was not revealed to the public until three hundred years after it was written, when it was rediscovered in 1908 in the archives of the Royal Library of Denmark.

Cristóbal de Albornoz (1530–1603) was a Spanish priest who fought idolatry in Peru in the sixteenth century, from 1568 until his death in 1603. He led a particularly intense campaign in the region of Huamanga (the current city of Ayacucho), which consisted in identifying and destroying temples frequented by the local population. In particular, he wrote an essay entitled *Instrucción para descubrir todas la guacas del Pirú y sus camayos y haziendas* [ca. 1582] (1984). Cristóbal de Albornoz recorded a total of thirty-seven temples in the *Chinchaysuyu* region, northwest of Cuzco, some not mentioned in the Cobo text.

Pedro Sarmiento de Gamboa (1532–1592) was a Spanish explorer, historian, and humanist. In the late 1560s, he placed himself in the service of Peru's fifth viceroy, Francisco de Toledo, who sought to consolidate Spanish rule over the country. It was in 1572, in Cuzco, that he wrote a *Historia de los Incas* [1572] (2007) which notably contained a detailed description of Inca mythology. He then organized several expeditions to the Strait of Magellan. The viceroy wanted to fortify this region because English ships were plundering Spanish boats on the Peruvian coast. After many adventures, he found himself a prisoner of the English before finally returning to Spain, where he died in 1592.

Miguel Cabello de Balboa (1535–1608) was a Spanish priest and writer. In 1566, he emigrated to Peru. Then he traveled to Quito, Ecuador, where he began writing his work entitled *Miscelánea antárctica, una historia del Perú antiguo*, finishing it in Lima in 1586 [1586] (2010). This work constitutes an important source of information on the northern Andes. The legendary origin of the Incas is presented in detail in his book, but his version relating to the historical origin of the Inca people differs somewhat from that of other Spanish authors.

Inca Garcilaso de la Vega (1539–1616), whose real name was Gómez Suárez de Figueroa, was a Spanish-speaking mixed-race chronicler born in Peru (in Cuzco). He was the son of the conquistador Sebastián Garcilaso de la Vega y Vargas and the Inca princess Isabel Chimpu Ocllo, who was a descendant of the Inca *Wayna Qhapaq*. He resided in Peru until 1560, then returned to Spain. He was the author of *Comentarios Reales de los Incas*, a work divided into two parts: the first concerned the history of his ancestors and the second the history of the conquest of the country. This is an interesting testimony because it is written by someone of non-European birth. However, this work was influenced by other writings because it appeared late [1609] (1969, 2000) (Fig. 3.3).

José de Acosta (ca. 1540–1600) was a Spanish Jesuit and missionary in South America. As a member of the Society of Jesus, he left Spain in 1569 for Lima in Peru, where he occupied a chair in theology. In 1571, he went to Cuzco,

PRIMERA PARTE DE LOS
COMMENTARIOS
REALES,
QVE TRATAN DEL ORI-
GEN DE LOS YNCAS, REYES QVE FVE-
RON DEL PERV, DE SV IDOLATRIA, LEYES, Y
gouierno en paz y en guerra: de fus vidas y con-
quiftas, y de todo lo que fue aquel Imperio y
fu Republica, antes que los Efpaño-
les paffaran a el.

*Efcritos por el Ynca Garcilaffo de la Vega, natural del Cozco,
y Capitan de fu Mageftad.*

DIRIGIDOS A LA SERENISSIMA PRIN-
cefa Doña Catalina de Portugal, Duqueza
de Barganca, &c.

Con licencia de la Sancta Inquificion, Ordinario, y Paço.
EN LISBOA:
En la officina de Pedro Crasbeeck.
Año de M. DCIX.

Fig. 3.3 Frontispiece of the first part of *Commentarios reales* by Garcilaso de la Vega, published in 1609. Biblioteca Nacional del Perú, public domain

where a Jesuit college had just been created. He founded various establishments and acquired a detailed knowledge of the country and its inhabitants. He then went to Mexico where he gathered a lot of information about this country and the Aztecs. He is best known for his work *Historia natural y moral de las Indias*, which appeared in Seville [1590] (1954). In this book, he proved himself to be a fine connoisseur of the peoples of North America (Aztecs) and South America (Incas). He derived much of his information from Polo de Ondegardo and, for this reason, cannot be considered an independent source.

Blas Valera (1545–1597) was born in Peru and studied in Trujillo, then in Lima. He joined the Society of Jesus and specialized in the study of local dialects, including Quechua. He wrote in an elegant, clear style and took care to carefully check all the information he provided. Among his writings, we find the text *Relación de las Costumbres antiguas de los Naturales del Pirú* (ca. 1585)(1950, 2009). He had problems with the Jesuit order and was imprisoned for several years. He returned to Cadiz in 1596 and died the following year, but the date of his death is controversial.

Diego González Holguín (1560–ca.1620) was a Spanish Jesuit and missionary who also established himself as a specialist in the Quechua language during the time of the viceroyalty of Peru. He arrived in this country in 1581 and published a grammar of classical Quechua, a dialect spoken at the Inca court. In 1608, he published a dictionary *Vocabulario de la lengua general de todo el Perú llamada lengua Quichua*, and in the same year, he also published *Privilegios concedidos à los indios* [1608] (1989).

Pablo José Arriaga (1564–1622) was a Spanish Jesuit and a missionary in South America. He was notably rector of the College of San Martin in Lima and rector of the College of Arequipa. It was in 1621 that he published his work *Extirpacion de la Idolatría del Pirú* [1621] (1968, 2012). The following year he left for Europe, but his ship was struck by a storm and he died in the shipwreck.

Antonio Vázquez de Espinosa (1570–1630) was a Spanish monk of the Order of Carmel. He visited many places in Peru during his stay there between 1615 and 1619. In particular, he stayed in Chavín, Cajamarca, and Arequipa. He was the author of *Compendio y descripción de las Indias Occidentales* [1628] (1942), a work which has become a reference on the history of Latin America since its discovery in the Vatican Apostolic Library in 1929. He was inspired by previous authors, notably Garcilaso de la Vega.

Giovanni Anello Oliva (1574–1642), known as Father Oliva, was born in Naples, Italy. An Italian Jesuit and historian, he died in Lima in 1642. He arrived in Peru in 1597 and entered the orders in 1601. He wrote a *Historia del reyno y provincias del Perú y varones insignes en santidad de la Compañia de Jesus* [1614] (1998). He contributed to the evangelization of the indigenous peoples and taught in different places (Chuquisaca, Cochabamba, Arequipa, etc.).

Fernando de Avendaño (ca. 1577–1665) (or Fernando Avendaño) was a Catholic priest who was born in Lima and lived in different places in Peru. He was notably rector of the University of San Marcos. He left notes on the primitive customs and rites of the indigenous peoples of Peru which are found in part in the writings of Pablo José Arriaga. His text *Sermones de los misterios de nuestra santa Fe católica en lengua castellana y la general del Inca* [1648] (2019) was published on the order of the Archbishop of Lima. These sermons were written in Quechua with a translation into Spanish.

Bernabé Cobo (1582–1657) was a Spanish Jesuit, writer, and naturalist. His main work is entitled *Historia del Nuevo Mundo* [1653] (1956). In the fourteen books of the first part of this work, he deals with the territories of the New World and what could be found there. The second part includes fifteen books devoted to the history of Peru, while the third part concerns the history of Mexico and neighboring territories. Only part of this work has reached us, the original edition being kept in the library of the University of Seville. The writings relating to Peru are inspired by the texts of Pedro Pizarro, enriched with personal research in the archives of the time. Part of Cobo's work derives from Polo de Ondegardo's initial text entitled *Los errores y supersticiones de los Indios*, mentioned above, which is now lost, but which contained more information than the later version published in 1585.

Antonio de la Calancha (1584–1654) was a friar of the order of Saint Augustine and an anthropologist of Bolivian origin who studied the indigenous peoples of South America. He studied in Lima and stayed notably in Cuzco. He obtained promotions in the order to which he belonged, which allowed him to travel to Peru during the viceroyalty. He published a work entitled *Corónica moralizada del orden de San Agustin en el Perú* [1638] (1981). He continued to collect information to write a second volume but did not finish it. It was his disciple, Brother Bernardo de Torres, who completed it and published it in 1655. These works contain a lot of information about the religion, customs, and traditions of the indigenous peoples of Peru and Bolivia.

Fernando de Montesinos (1593–1655) was a Spanish writer, historian, and ecclesiastic who traveled to Peru from 1628. He was the author of *Memorias antiguas, historiales y politicas del Perú* [1630] (1882, 1906). The credibility of this work is questionable, particularly with regard to the Inca dynasty. He also wrote *Anales del Perú, 1498–1642*, which appeared in Madrid in two volumes in 1906. It is a more trustworthy work than the previous one, dealing with the successes of the conquest of the New World and colonization, which began with the third voyage of Christopher Columbus in 1498.

Juan de Santa Cruz Pachakuteq Yanki Salqamaywa was born in Peru at the end of the sixteenth century. He was a Peruvian chronicler, descendant of the local Aymara nobility and author of *Relación de las antiguedades desde Reyno del Pirú*, a very interesting source for its ethno-historical content. It was probably written in 1613 (or a little later, between 1620 and 1630) but was only published in 1950 by Marcos Jiménez de la Espada with two other chronicles under the title *Tres relaciones de antiguedades peruanas*. The original manuscript is in the National Library of Madrid. Juan de Santa Cruz Pachakuteq Yanki Salqamaywa was the author of a representation of the Inca cosmovision or world view in the form of a drawing found in the Temple of the Sun in Cuzco (see Sect. 10.17).

Luis de Monzón (dates unknown) was a magistrate who accompanied the Spanish conquistadors at the end of the sixteenth century. He was the author of a work entitled *Descripción de la tierra del repartimiento de los Rucanas Antamarcas de la Corona Real, jurisdicción de la ciudad de Guamanga, año de 1586* (1881).

Juan de Velasco y Pérez Petroche (1727–1792) was an Ecuadorian historian and Jesuit who taught philosophy and theology at the Real Audiencia of Quito. We owe him in particular a work entitled *Historia moderna del Reino de Quito y crónica de la provincia de la Compañía* [1789] (1979).

A text entitled *Discurso de la sucesión y gobierno de los Yngas* is attributed to an anonymous chronicler [ca. 1570] (1906).

The Huarochiri manuscript (1991) is a text in the Quechua language from the end of the sixteenth century or the beginning of the seventeenth century which describes the myths, religious beliefs, and traditions of the indigenous peoples in the province of Huarochiri. It gives pride of place to mountain deities considered to be protectors of local ethnic groups. The document, whose author is unknown, was annotated by the Spanish clergyman Francisco de Avila (1573–1647), who was responsible for the eradication of pagan beliefs in the Andes. The manuscript was rediscovered in the twentieth century at the Royal Library of Madrid (Salomon & Urioste 1991).

3.4 The *Khipukamayoqs* or 'Khipu Masters'

The Incas did not have a writing system in the sense we understand it today, but they used ingenious devices to convey information in a fairly convenient way (Ascher & Ascher 1981, 1997). These were the *khipus*, sets of cords used to represent numbers in the decimal system (Fig. 3.4). Note that, in contrast to the Mayans, who used a vigesimal system (base twenty) to count, the Incas used the decimal system so familiar to us today. *Khipu* means 'knot' or 'count' in Quechua.

Khipus were used by the Inca administration to count statistical data relating to the economy or the organization of society (Radicati di Primeglio 1976; de Pasquale 2011). Making a *khipu* required having a horizontal rope (the main rope), hanging cords vertically (the secondary ropes) to which one could attach still more ropes (subsidiary ropes). The information was translated in the form of nodes arranged on the suspended cords (secondary and subsidiary). *Khipus* were made using cotton or camelid wool (particularly alpaca), but they could also contain plant fibers. Different colors were used, and we have even found cords whose color changed mid-length (Hyland 2017).

Fig. 3.4 The *khipus* were sets of knotted cords for conveying information in the decimal system. Author's photograph

Numbers were represented by sequences of knots on the secondary cords. Bigger and bigger knots were used to record the numbers 1–9. From the end of the secondary rope, each knot counted the units, then the tens, the hundreds, the thousands, and so on, with zero being represented by the absence of a knot. Garcilaso de la Vega [1609] (1969, 2000) writes:

> Depending on their position, the knots represented units, tens, hundreds, thousands, tens of thousands, and exceptionally, hundreds of thousands, and they are all well aligned on their different strings, like the numbers that an accountant would enter, column by column, in his ledger.

It may be that some *khipus* conveyed other information than simple lists of numbers relating to tributes, namely real texts or narrations, and a wide variety of data such as administrative (taxes, census), genealogical, calendar, and historical information, and perhaps more still. These messages could have been translated by the choice of the type of material used (cotton or wool), by the use of different colors, by the particular twist of the fibers, or by the appearance (front or back) of ligatures. It seems that various aspects of the law could also be expressed via devices of this type, and that these notably contained the sentences applicable to different offenses.

Khipus were not invented by the Incas, but were used to record data by earlier civilizations which flourished in what is now Peru. They are known, however, thanks to the intensive use made of them by the Inca administration, for the social and economic management of their empire. In each community, there was a specialist to manage these devices called the *khipukamayoq*, or *khipu* master, who could read off the relevant digital data. Important information was encoded and used to monitor the prosperity of the different communities (Pareja 1986).

Some chroniclers like Blas Valera [ca. 1585] (1950, 2009), Miguel Cabello de Balboa [1586] (1951, 2010), and Giovanni Anello Oliva [1614] (1998) mention the fact that they received information from the *khipukamayoqs*. If we are to believe Bernabé Cobo [1653] (1956) and Father Martin de Murúa [ca. 1615] (2008), many ancient facts were recorded by these rather particular kinds of secretary, and there were different *khipus* for tributes, for statistics valid for different territories, for rites and ceremonies, and indeed other areas. Instruction passed from generation to generation, and those who were chosen to carry out this task were not necessarily capable of understanding the registers and notes of other officials. Father Martin de Murúa writes that he was very surprised that knotted strings could tell of so many things relating to the past, including the length of the reign of each Inca king, whether he was brave or not, in short "everything that one generally finds in books." We have not yet been able to decipher the codes used in these so-called 'historical' *khipus*, which retrace the history of the Inca dynasties and were probably used for mnemonic purposes.

The *khipus* were transported by *chaskis*, who traveled on foot through the different regions of the empire (Fig. 3.5). This was a very effective way to transmit quantitative but also possibly qualitative data from one place to another quite quickly in the form of coded messages.

Some authors have noticed the similarity between the structures of the *seq'es* and the *khipus*. *Khipus* have been studied in detail by Urton (1998, 2003, 2005), Quilter & Urton (2002). It is important to mention that many *khipus*, like the codices, were destroyed by the Spanish conquerors in their fight against idolatry, but several hundred remain.

Khipus were also used to store information of an astronomical nature, as evidenced by an illustration taken from the *Nuova Cronica y Buen Gobierno* by Guamán Poma de Ayala [1615] (1936, 1980), which represents an astrologer holding such an device in his hand and who is assumed to be observing the celestial bodies in order to determine the season for planting cereals and potatoes in the fields.

Fig. 3.5 The chaskis wore a particular livery specific to their profession. They were chosen for their loyalty to the emperor and their sporting abilities. Bolivian postage stamp

Most of the *khipus* we have today were found in tombs, which may mean that the people in these tombs were *khipukamayoqs*, because it was common to bury the deceased with objects that belonged to them, so that they could take these things with them on their journey to the next world.

As recently as 2013, *khipus* were found at the archaeological site of Incahuasi, in the Cañete Valley, 160 km south of Lima. They were discovered in buildings where agricultural products were stored and were undoubtedly used as inventories of foodstuffs such as peanuts, peppers, beans, corn, etc. Given the extreme drought conditions in this area, these *khipus* were well preserved. One of the *khipus* had its knots undone, suggesting that information had been removed and that such a device could be reused several times.

To date, scientists have not been able to decipher these *khipus*. Among around 900 objects of this type so far identified, around a third appear to contain more elaborate messages than the others. This impression results from the choice of colors of the cords and the way in which they are arranged, the structure and orientation of the knots, and the way in which the cords are attached to the main rope. Given the number of possible combinations, this may mean that they constitute a form of writing, for example of syllabic type, but this is only a hypothesis. If one day we manage to unravel their secrets, these devices may provide us with much interesting and useful information about the Inca empire.

4
History of Andean Civilizations

4.1 Chronology and Emergence of the Different Civilizations

At its apogee, the Inca empire was immense. Its power extended over a strip of territory more than 4000 km long, covering a major part of the current states of Ecuador, Peru, Chile, and certain regions of Argentina and Bolivia. The influence and prestige of this amazing civilization was felt far beyond the limits of the conquered territories and even reached Panama and Brazil. It is interesting to retrace the main stages of the formation of this very real and geographically extensive military power, bearing in mind that the empire that the Spaniards found when they arrived in these regions was in fact fairly recent, since the imperial period only dates back to the fifteenth century. This civilization, however, was largely dependent, in certain respects, on the artistic and architectural contributions of previous cultures which had flourished over the centuries in the different regions of present-day Peru (von Hagen & Le, 1979; Bauer, 1992).

The predecessors of the Incas in this part of the world did indeed contribute to the formation of a refined civilization, which the archives and archaeological finds are gradually revealing to us. To understand the Inca civilization, it is therefore essential to delve into the different cultures that preceded it, for without their contributions, this world that still captivates us would undoubtedly not have emerged at all. It is the aim of this chapter to recall some essential contributions of these different cultures. More details can be found in numerous works on the subject (see, for example, Mason, 1968; Bankes, 1977;

Lavallée & Lumbreras, 1985; Coe et al., 1987; Moseley, 1992; Longhena, 1999; Longhena & Alva 1999; Dupas, 2002).

Over the centuries, several important cultures emerged then declined in the three major ecosystems of the Andes: the tropical or Amazonian forest, the highlands, and the river valleys separating the often desert regions from the coast. Some peoples showed themselves to be more innovative and more inclined to conquest than others, but interactions between the different civilizations were inevitable and sometimes even occurred over long distances.

Where the prehistory and history of Andean civilizations are concerned, it is usual for archaeologists to divide these into seven major chronological periods. Three of these periods remain in history as eras of great cultural unity and are listed under the terms Early Horizon (from 1000 BCE to the beginning of the common era), Middle Horizon (650 to 1100 CE) and Recent Horizon (1450–1550 CE). The Ancient Horizon saw the emergence and domination of the civilizations of Chavín, Cupisnique, and Paracas, the Middle Horizon corresponds to the reigns of the Wari and Tiwanaku empires (Huari and Tiahuanaco, in Spanish), while the Recent Horizon saw the emergence and supremacy of the Inca empire until the arrival of the Spanish in 1532.

The Early Intermediate and Recent Intermediate periods saw the development of more complex and less homogeneous regional structures after the decline of the Chavín and Wari/Tiwanaku civilizations, namely the emergence and development of the Nazca and Mochica societies, on the one hand, and Chimú, Chancay, and Chincha, on the other (see Table 4.1 and Fig. 4.1).

Previously, the pre-ceramic period had seen the arrival of people in these territories and the emergence of the first settled communities. The initial period is characterized by the appearance of pottery, the first woven textiles, and in another area, the birth and development of large ceremonial centers.

It would not be wrong to assert that the pinnacle of Andean civilization was achieved by the Inca empire, which succeeded in a very short time in conquering the territories where the previous civilizations had flourished. From their capital Cuzco, located in the southern highlands of the central Andes, the Incas extended their dominion south to the Chilean desert, to northeastern Argentina, east to the Bolivian altiplano, and to the north and west of present-day Peru, up to and including Ecuador. It was a gigantic territory with landscapes of extraordinary variety, including almost impenetrable forests, inhospitable deserts, mountains with very high peaks and deep valleys, fertile river areas, and a very extensive marine coastline. It was by making the best use of the varied resources of these harsh and often inhospitable lands that the various Andean civilizations had emerged and grown.

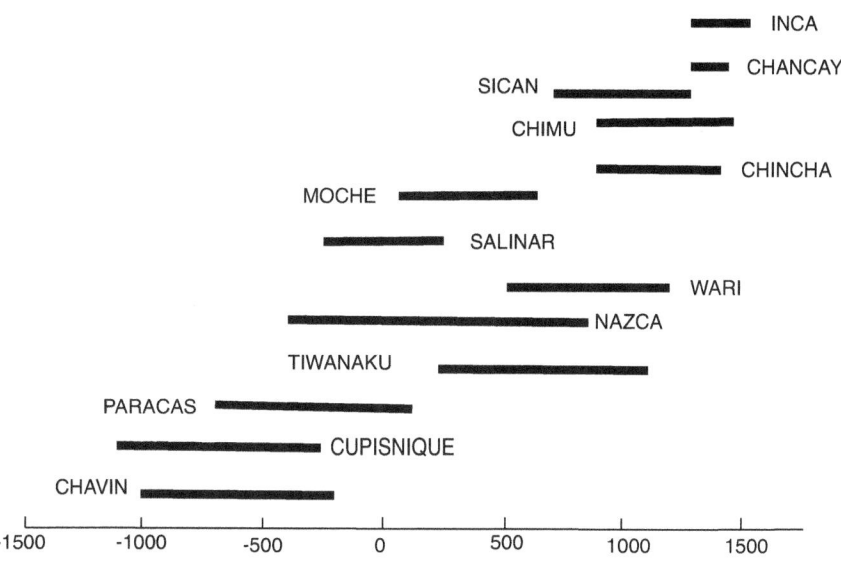

Fig. 4.1 Panorama and chronology of the main Andean cultures discussed in this text. Author's drawing

Table 4.1 summarizes the fundamental chronology of the main andean civilizations.

Table 4.1 Chronology of the main Andean civilizations. Adapted from Cavatrunci et al. (2005). The beginning of the Chimú and Chancay cultures is consistent with Itier (2008)

Period	Approximate duration	Civilization
Pre-ceramic period	30th–19th century BCE	
Initial period	19th–10th century BCE	
Ancient horizon	10th–2nd century BCE	Chavín
	9th–2nd century BCE	Cupisnique
	4th century BCE–2nd century CE	Paracas
Ancient intermediate period	3rd century BCE–6th century CE	Nazca
	3rd century BCE–9th century CE	Mochica
Average horizon	6th–10th century CE	Wari
	7th–13th century CE	Tiwanaku
Recent intermediate period	11th–15th century CE	Chimú
	11th–15th century CE	Chancay
	14th–15th century CE	Chincha
Recent horizon	15th–16th century CE	Inca

4.2 Chavín de Huántar and the Lanzón

Around 900 BCE, commercial exchanges were very intense throughout the area of Peru between Cajamarca in the north and Ayacucho in the south. In this context, a major ceremonial center emerged, generating the so-called Chavín civilization. Chavín de Huántar was an important locality, located at almost 3200 m above sea level. It controlled trade routes east to the Amazon and west to the Pacific. This site was located at the confluence of two rivers (the Huacheksa and the Mosna) in the Marañón River basin. This civilization would spread effigies of the 'Staff God,' a centerpiece of its religious iconography, throughout a large part of Peru (Burger, 1992). The influence of this place was considerable for several centuries. The ceremonial center was very extensive. Indeed, a whole complex of underground drainage canals and residential settlements have been discovered in the surrounding area.

The religious and administrative center of the Chavín culture included structures in the shape of truncated pyramids, the most imposing of which bears the name 'El Castillo' (or 'Templo Mayor') attributed to it by the Spaniards. It achieved a high degree of perfection, through both its architecture and the litho-sculpture associated with it. It was an important pilgrimage center in the Andean world. The Peruvian archaeologist and anthropologist Julio César Tello Rojas (1880–1947) promoted this site in the period after the 1920s as the cradle of Andean culture (see Sect. 3.1). The 'Old Temple' ('Templo Viejo') had a U-shaped structure with a central courtyard. On this site, there are obelisks and monoliths with sculptures representing caimans, jaguars, and falcons, and also anthropomorphic representations. The best known monoliths are the Lanzón ('the great spear'), the Raimondi stele,[1] and the Tello obelisk, which had an eminently sacred character. The Lanzón shows an anthropomorphic god with the head of a feline and hair in the form of snakes. He has a fierce look and a mouth full of fangs. This monolith was associated with the cult of fertility, and paying homage to it guaranteed bountiful harvests.

The site of Chavín was run by priests and priestesses who held considerable power through their detailed observation of the stars and agrarian cycles. These priests were keen observers of the sky and were attentive to the movements of the Sun, the phases of the Moon, and the apparent motions of the stars. Knowing the details of agrarian cycles, they enjoyed a very high social status. Chavín de Huántar was frequented by many pilgrims in search of oracles. It is believed that the use of hallucinogenic substances was common during the magico-religious rites that were celebrated there.

---

[1] Antonio Raimondi (1826–1890) was an Italian explorer, naturalist, geographer, and botanist who published many works on Peru.

The characteristic art of the Chavín culture, which flourished during the first millennium BCE, is found in architecture, sculpture, ceramics, and even textiles (Domenici, 2009). In lithic art, stones are engraved or bear designs in relief. Buildings and their annexes have columns, cornices, tenon heads, and lintels, and there are also tombstones and obelisks. There are sculptures on the walls, and ceremonial spaces are decorated with the gods and demons of the local pantheon (Barbier, 1999). Chavín ceramics are often shiny black. They have abstract shapes but also include zoomorphic symbols of the supreme divinity. Ceramics from other places were found at the Chavín site, indicating that pilgrims had traveled from very distant regions to pay homage to local deities.

The creative genius of the Chavín artists is attested by their ability to achieve a synthesis of the symbolism of the coast, the highlands, and the rainforest. We do indeed find an Amazonian influence in the iconography, which represents a bestiary specific to the forest (caimans, jaguars, pumas, snakes). The image of the jaguar, which crystallizes a cult specific to this civilization, spread rapidly from 900 BCE, from Pichiche (in the north) to Ayacucho (in the south), and indicates the political, cultural, and religious hegemony of the priests of Chavín, who greatly contributed to unifying a large part of pre-Hispanic Peru.

Archaeologists differ over the chronology of the site and the Chavín culture. According to a long chronology, the first buildings date from 1300 BCE, the peak to around 800 BCE, and the abandonment of the ceremonial center to around 400 BCE. However, according to a short chronology based on the analysis of ceramics, sculptures, etc., the first buildings date back to 850 BCE, while the influence of the site would have reached its peak around 500 BCE, and the decline, with the abandonment of the temples, would have occurred around 200 BCE.

No one knows whether the Chavín civilization experienced a slow decline or collapsed quickly. Very severe climatic episodes, such as heavy rains associated with *El Niño*, could have played a role. But for the cult of Chavín to disappear, it undoubtedly required major cultural and social tensions, possibly reinforced by a large-scale ecological disaster.

4.3 Lambayeque and Cupisnique Culture

Cupisnique culture was partly contemporary with that of Chavín, reaching its peak from around 900 to 200 BCE. It spread especially in the northern coastal valleys of Lambayeque and Plura (Fig. 4.2). It is now recognized as having independent characteristics, but it was long considered to be a stylistic varia-

Fig. 4.2 Vase with stirrup handle dating from the Cupisnique period. The stirrup handles remained a characteristic of this region until the Inca domination. Author's photograph taken at the Larco museum in Lima

tion of Chavín culture. It was the Peruvian archaeologist Rafael Larco Hoyle (1901–1966) who first distinguished the Cupisnique and Chavín cultures (see Sect. 3.1).

Cupisnique culture was one of the first to produce high quality ceramics. The potters popularized the stirrup-handled bottle, and it remained characteristic of this region until the Inca domination. The ceramic is usually black and shiny, and sometimes brown, but always monochrome. The misleadingly massive aspect of these artistic creations sometimes gives them the appearance of basalt or obsidian (Fig. 4.3). The inspiration comes from the plant and animal world, frequently with a feline theme, but is also influenced by a religious pantheon close to that of Chavín.

When the Chavín de Huántar culture began to decline, small fiefdoms on the north coast came together to form the Salinar culture, which began between 500 BCE and 200 BCE and ended around 300 CE. This culture established

Fig. 4.3 Example of a monochrome black ceramic showing a vase with a stirrup handle in the shape of a dog. Author's photograph taken at the Larco museum in Lima

the transition between the Cupisnique and Chavín cultures and the advent of the Mochica civilization.

4.4 Paracas Culture and the *Fardos*

After the decline of Chavín, new cultural forms appeared along the southern coast of Peru. From 800 BCE to 200 CE, we see the flowering of a style of pottery called Paracas, named after the peninsula located approximately 250 km south of Lima. The Paracas potters did not use the wheel and paint was applied after firing. The lines they engraved were filled with dense paint in different shades (red, yellow, blue, green, etc.), and the patterns were often abstract and geometric. Towards the end of the Paracas period, lighter shades were adopted and potters also began to use slip.[2]

The region of the Paracas peninsula was the seat of major funeral centers. It was the archaeologist Julio César Tello Rojas who uncovered, in 1925, two necropolises where tombs containing numerous very high quality fabrics

---

[2] Slip is a colored coating applied to ceramics before firing to hide the natural color.

were discovered. In fact, two types of tombs were brought to light, apparently reflecting different periods. On the one hand, there were the *cavernas* ('caves'), consisting of underground burial chambers accessible by a vertical shaft, and each containing several bodies (without doubt of the same clan or the same family), accompanied by offerings. And on the other hand, there were the *necropolis* ('necropolises') with square masonry chambers containing the *fardos*. These were mummies wrapped in several very high quality fabrics with rich decorations. Depending on the social level of the deceased, the mummies were honored with multiple embroidered and polychrome fabrics including large *mantos*, with various accessories such as pins, necklaces, or scepters, which were symbols of power.

The decorative patterns of the fabrics refer in particular to fertility rites, which is not surprising in these arid regions, but they also evoke severed heads, which were often war trophies. The Paracas civilization practised cranial deformation, undoubtedly for aesthetic purposes, as well as ritual trepanation. The decorations of the fabrics and pottery suggest that the inhabitants of these regions were followers of a complex cosmovision which anticipated the Nazca culture.

4.5 Nazca Culture and Mummified Heads

The Nazca culture developed in Peru during the period between approximately 200 BCE and 600 CE, or during the Early intermediate period. It flourished on the southern coast and represented, in this region, a very high cultural level. It developed in parallel with the Mochica (or Moche) civilization which appeared on the northern coast. Its history is characterized by several phases driven by small centers with an essentially agricultural vocation. An important city of this civilization was Cahuachi, which was the ceremonial center of reference for all the populations of this region.

The Nazca civilization expanded into the semi-desert area located between the Pacific Ocean and the Andes mountains. Its sphere of influence corresponded to the current department of Ica, the center being the Rio Grande valley in the province of Nazca about 400 km south of Lima. Among the characteristics of this culture, we find the development of very rich ceramics, the manufacture of particularly sumptuous textiles, possibly decorated with bird feathers, and the use of trophy heads. Another exceptional feature are the spectacular geoglyphs carved into the desert pampas of the coast, discussed in the following section.

The Nazcas practiced intensive agriculture in the valleys along the tributaries of the Rio Grande, as well as in the Ica Valley. As this region is particularly arid, they were forced to develop quite remarkable irrigation, including a network of underground aqueducts.[3] The people lived in small houses with thatched roofs, grouped into villages located on the borders of cultivated territories. In the ceremonial center of Cahuachi, not far from the current city of Nazca, the ruins of imposing official buildings built from adobe have been found, along with tall step pyramids serving as sanctuaries. The Nazcas practised fishing and sealing, and they cultivated cotton, beans, potatoes, cassava, avocados, and peanuts.

Nazca ceramics are truly spectacular for the symbolism of their motifs and the richness of their polychromy. Influenced by the Paracas culture, it was not initially very colorful, but it then went on to use six colors—white, black, gray, red, orange, and brown—and this eventually grew to a dozen. At first, it favored natural forms such as fruit, plants, animals, and human forms, displaying also religious and mythological themes, before developing a more abstract iconography. Pottery decorations indicate that these people glorified animal deities, including felines, birds, and snakes. The ceramics found at Cahuachi appear to have had a religious vocation, and it is after the decline of this cult center that we begin to see decorations featuring faces and human forms (Barbier, 1999; Domenici, 2009) (Figs. 4.4, 4.5, and 4.6).

Fig. 4.4 The first phase of the Nazca civilization was contemporary with the Paracas civilization. On this double vase, a Nazca warrior teaches his son the 'art' of decapitation. Author's photograph taken at the Larco museum in Lima

---

[3] This system is similar to the foggaras of North Africa, which are draining galleries under the beds of intermittent rivers.

Fig. 4.5 Vase with an anthropomorphic shape characteristic of the Nazca culture. Author's photograph taken at the Larco museum in Lima

The Nazcas made textiles of great beauty using cotton and wool woven from llamas and alpacas. Influenced first by the Paracas civilization, their textile art incorporated wefts and warps as well as repetitive themes. They then sought more complexity in their designs. They introduced iconography, including mythical beings, added embroidery, and developed geometric and abstract representations. Very beautiful coats intended for high political and religious authorities were made using feathers.

One Nazca custom consisted in mummifying human heads (men, women, and even children) by sewing their lips with thorns. These heads, often belonging to enemy soldiers or their families, were worn suspended on ropes, the idea being to benefit from the courage and strength of the deceased. Undoubtedly for aesthetic purposes, they also practised deformation of the skull by requiring newborns to wear leather straps.

Fig. 4.6 Ceramics from the Nazca period. The ceramics of this culture are particularly spectacular for the symbolism of their motifs and the richness of their polychromy. Author's photograph taken at the Larco museum in Lima

The Nazca civilization declined after 350 CE. It is assumed that this was a consequence of some very violent natural phenomena such as an earthquake, a major flood, or even a severe drought. Major floods caused by exceptional *El Niño* phenomena would have had a devastating effect on this agricultural civilization. Traces of the Nazca culture went on for some time before it was assimilated by the Wari civilization.

4.6 The Mystery of the Nazca Geoglyphs

The Diversity of the Geoglyphs

A particularly fascinating and mysterious feature of the Nazca culture are the well-known geoglyphs, often very large, traced on the desert floor. An overall view is only accessible from a certain height. The ground where these figures are drawn is a carpet of dust and stones covered with reddish traces of iron oxide. By scraping through the surface, the Nazcas revealed a characteristic grayish gypsum soil which clearly marks out the figures.

The geoglyphs are distributed geographically along a line connecting the towns of Nazca and Palpa in the Ica region, with the main concentration being located in a rectangle of around forty square kilometers south of the village of San Miguel de la Pascana. Travelers who crossed the plains of San José had long known of the existence of mysterious lines in these regions. These so-called 'Nazca lines' were 'rediscovered' in 1927 by the Peruvian archaeologist Toribio Mejia. That year, during the winter solstice in the southern hemisphere, he noticed that the Sun was setting along one of these lines. He deduced that it had been drawn as a solstice line for the agricultural calendar of the ancient peoples who inhabited the region, and he concluded that it thus constituted 'the largest astronomical calendar in the world' (Kosok, 1965) (Figs. 4.7 and 4.8).

In addition to the geoglyphs of the Nazca pampas, so often studied and publicized in the past, other figures are traced on the ground near the town of Palpa and not far from the Rio Grande, the Rio Palpa, and the Rio Viscas. A detailed study of the Palpa geoglyphs has been carried out by Lambers (2004).

The study of the Nazca geoglyphs is inseparable from the name of Maria Reiche (1903–1998), a German mathematician and archaeologist, who was born in Dresden and died in Lima. She spent the major part of her life studying and preserving these figures, considering them to have an astronomical function

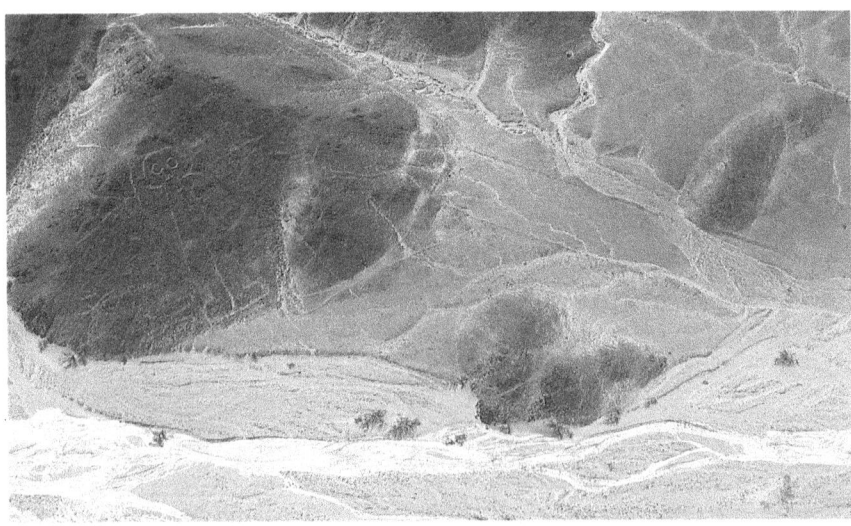

Fig. 4.7 Example of a geoglyph from the Nazca region. We see the 'cosmonaut' in the left part of the photo. Author's photograph

Fig. 4.8 Another example of a geoglyph showing geometric shapes. Author's photograph

linked to agriculture (Reiche, 1968; Kosok & Reiche, 1947). Since then, many researchers have carried out on-site studies and many new geoglyphs, some of small dimensions (a few meters to a few tens of meters), have been discovered in recent decades. The dating of the geoglyphs, using carbon-14, has shown that a certain number of them predate and others postdate the Nazca civilization. Some figures discovered in 2018 using satellite and drone data date back to the Paracas culture.

Hundreds of geometric figures have been identified, sometimes very large– several hundred meters or even several kilometers across. They are made up of straight lines, spirals, ellipses, trapezoids, and triangles. They have anthropomorphic forms ('the man with the head of an owl,' 'the astronaut,' etc.), or zoomorphic forms (monkey, hummingbird, condor, jaguar, spider, orca, pelican, etc.), and even phytomorphic forms with representations of stylized plants. Tracing them out did not require the use of advanced technologies, only simple geometric processes involving wooden stakes, ropes, the use of grids, and the like. Indeed, it was a straightforward matter to clear the surface stones burned by the Sun and create a furrow of a different color, given the gypsum-rich soil beneath, bordered by pebbles.

The excellent preservation of these figures can be put down to several factors: firstly, the particularly dry climate of these regions, where it rains extremely rarely, then the configuration of the terrain with winds blowing primarily from south to north making it impossible for sand to accumulate on the flat surface, not to mention the high air temperature at ground level which protects the geoglyphs from the wind and the fact that the gypsum in the soil creates a form of plaster which, combined with the low local humidity, retains the dust.

It is interesting to note that many of the figures represented correspond to animal deities from the Nazca pantheon which are frequently found on pottery in similar or stylized forms. Birds played a major role in the local mythology: this was the case of the hummingbird, a typical bird in this region, and the condor, also seen in the area (Fig. 4.9). The frigatebird, on the other hand, represented several times, is rarely seen in this region. The presence of marine iconography (fish, orcas, whales) in the geoglyphs is linked to the fact that the Nazca civilization was established in coastal areas and that sailors were able to observe these creatures on the high seas. This is why they are also depicted on pottery and fabrics.

Fig. 4.9 The king of Peruvian birds of prey is the Andean condor, which can have a wingspan of three meters. Author's photograph

The Mystery Remains

Many hypotheses have been formulated to give a coherent explanation for these geoglyphs. Some are quite rational, others more daring or even decidedly fanciful, and we shall not be concerned with the latter here. However, no definitive explanation has been offered to date.

According to some (Kosok & Reiche, 1947), these figures and lines are linked to the observation of astronomical phenomena such as the rising and setting of the Sun at the solstices and equinoxes, the heliacal rising and setting of certain stars like Canopus, Sirius, or Capella. Canopus (or α Carinae) is known to be the brightest star in the southern constellation Carina and the second brightest star in the night sky after Sirius (α Canis Majoris) (see Appendix C). Capella (or α Aurigae) is the brightest star in the constellation of the Charioteer, and easily observable, like Sirius, Arcturus, and Vega. One possibility could be that the Nazca figures are representations on the ground of constellations observed in the sky.

The geoglyphs may actually be part of an immense calendar, in fact designed to work as an agricultural calendar. The key dates determined from this calendar would have had a religious meaning and would have been associated with particular rites and sacrifices. The grid-shaped drawings may have related to the Moon, constituting some kind of enormous device for anticipating the lunar cycles. According to Maria Reiche, the Nazcas' astronomical observations could have been recorded in this way for prosperity. This would explain why they engraved them in the pampas, so that they could be passed on to their descendants.

According to the same author, the Nazcas considered the constellations as deities who created natural phenomena. The idea was that one could exert some influence on them through a cult, notably through sacrifices. The figures on the pampas could therefore represent both a large astronomical calendar and a temple intended for ritual. However, such theories have been the subject of controversy because, according to their detractors, it cannot be demonstrated that all the observed figures are linked to the planets or stars and their constellations. Furthermore, these figures do not exhibit mathematical relationships, so cannot be extrapolated to astronomical studies and the observation of stellar asterisms.

For some researchers, the drawings represent dancing figures of a sacred nature, as part of a cult devoted to the ancestors. The celebrants would then follow in a procession along the paths traced on the ground. For still others, the geoglyphs could be linked to a cult dedicated to water and fertility in a region where drought was a permanent feature and where it was considered important to invoke the associated deities in order to survive in the desert. The figures could also correspond to representations of deities of this type as they were visualized by shamans following the consumption of hallucinogenic plants, but this explanation is implausible, even though the consumption of hallucinatory substances in ancient Peru was no exception.

Regarding the cult of water and fertility, it is interesting to note that straight lines had a sacred character for most Andean cultures, and not only the Nazcas. Their role would have been to connect sacred spaces like the mountain peaks which were believed to contain the spirit of a god with some influence over the climate, and therefore over rain, such as would be vital to the survival of the people. In the case of the Nazca lines, the three mountains associated with fertility myths were called Cerro Blanco, Illa-Kata, and Tunga. It has also been shown that certain lines correspond to underground channels where water flows, or indicate points of intersection of such channels. Although it almost never rains in the Nazca region, it is worth remembering that water travels underground from the Andes mountain range to the ocean. The canals which transported water to reservoirs were between 3 and 6 m below the surface. They were covered with flat stones and had ventilation holes which were used for cleaning work.

The figures in the form of zigzags and spirals may also have been associated with the cult of water, with the zigzags representing lightning and the spirals corresponding to the ventilation holes dug along the irrigation canals, which also had this shape. The spirals may even have evoked the seashells found in abundance in these regions.

Over the past decades, new hypotheses have been formulated on the basis of archaeological, anthropological, and ethnohistorical studies. Some researchers have tried to integrate geoglyphs into a broader context and interpret them by evoking Andean traditions involving the social organization, religious practices, and cultural concepts of the local populations. The geoglyphs would have been created by a population organized into social groups whose members shared common ancestors. These groups would have met in the desert, marking out a common space according to beliefs deeply rooted in their traditions. This social interaction was important to each group's position in a broader societal context.

A cult centered around mountain deities, water flow, and fertility, organized according to a religious calendar, would undoubtedly have been the basis of geoglyph activity. The geoglyphs would have been connected to sacred places and used in ritual processions by the groups who created and maintained them. Ceramic vessels, believed to be filled with food and drink, were ritually smashed and placed on and along the centers of the lines. Certain geometric figures could have marked places where larger groups gathered, while biomorphic figures, whose patterns evoked the concept of fertility, would have been used for dances. All in all, the geoglyphs would have been deeply rooted in the daily lives of the Nazca people, and the basic concepts underlying their creation would have been in accord with cultural, religious, and social traditions.

For a more detailed discussion and numerous references on the subject, we refer to the work of A. F. Aveni entitled *Between the Lines. The Mystery of the Giant Ground Drawings of Ancient Nazca* (2000).

4.7 The Mochica Theocracy

Following the waning influence of Chavín de Huántar, the Mochica (or Moche) culture flourished on the northern coast of Peru roughly between 100 and 700 CE. It was contemporary with the Nazca civilization which flourished on the southern coast, and arose with the unification of the kingdom of Vicús in the north near the Ecuadorian border, the Salinar culture in the Chicama valley, and the culture of Virú thriving further south than the kingdom of Salinar. Two centers developed almost simultaneously, namely Moche and Sipán, in the Lambayeque valley, but the Moche ended up dominating the entire northern coast. To maintain the borders, they built impressive military fortresses at Pampa Grande in the north and Pañamarca in the south, with warfare playing an important role in this culture.

Mochica society was highly centralized and hierarchical. It is assumed to have been a theocratic state, with the sovereign also fulfilling priestly functions. Apart from its rulers, this society consisted of warriors, priests, artisans, fishermen, and farmers. It was opulent if we can judge by the scale of the temples and the luxuriance of the palaces built in adobe and decorated with polychrome frescoes.

The center of this culture was located in Moche, where there were two iconic stepped pyramid structures to which the Spanish gave the names *Huaca del Sol* and *Huaca de la Luna*, the center of religious ceremonies being the *Huaca de la Luna*. The main god (Ai Apaec) was called the 'decapitator' ('El degollador') because he is often represented holding a knife in his hand and a human head held up by the hair. This iconography is a very clear allusion to the human sacrifices which were part of religious rituals. Indeed, the Mochicas sacrificed human beings to influence the gods and in particular to try to avoid excessive rains, which could prove very destructive to their crops (see, for example, Bourget, 2016). They built irrigation canals and aqueducts, enabling intensive farming of corn, squash, and beans.

Fig. 4.10 Some examples of Mochica ceramics. Produced in large quantities, these ceramics are well known. Author's photograph taken at the Larco museum in Lima

Artisans mastered the metallurgy of copper, silver, and gold and produced alloys such as bronze to make weapons, tools, jewelry, and decorative objects (Domenici, 2009). They also inlaid precious metals with semi-precious stones (like lapis lazuli) and shells like spondyls (Figs. 4.10, 4.11, and 4.12).

Mochica ceramics were of great importance (Villacorta Ostolaza, 2007). Artisans used molds to produce them in large quantities. It was meticulous work. They favored certain colors, such as red or black on a cream background, which differed from the rather shimmering shades of the Nazca pottery. Illustrations depict animals or scenes from mythology or everyday life (agriculture, metallurgy, weaving, land hunting or sea fishing, etc.), including the representation of sexual relations (Hocquenghem, 1979) (Fig. 4.13). The iconography is also often dominated by ritualized violence, with numerous human sacrifices, those immolated being defeated warriors or sacrificial victims (Barbier, 1999). Human sacrifices were indeed frequent in this culture, some of them carried out on mountain tops.

Recent history has been marked by several major archaeological discoveries relating to the Moche civilization. The excavations carried out at the Huaca Rajada site in 1987 by the anthropologist Walter Alva and his wife Susana Meneses uncovered the now famous tomb of the 'Lord of Sipán' ('El Señor de Sipán') mentioned above (see Sect. 3.2). In 2008, archaeological excavations at the site of Huaca el Pueblo (Zaña valley) revealed the burial of a Mochica leader nicknamed 'El Señor de Ucupe.' His remains were accompanied by those of other people buried in a tomb containing several hundred funerary objects.

Fig. 4.11 Mochica portrait vase with stirrup handle representing a warrior with ornithomorphic headgear. Author's photograph taken at the Larco museum in Lima

We know that the mummies found in these tombs were those of high-ranking individuals because they were buried with servants, and the scenes depicted in the works of art that were found there illustrate mythological tales or rituals officiated by priests dressed as gods. The tombs of priestesses attest that women could reach high ranks in the religious hierarchy.

The Moche civilization, a rural society led by a powerful caste of warrior-priests, is believed to have permanently collapsed after the year 600 CE. What were the reasons for this? Among the prevailing theories, we can mention violent climate variations such as torrential rains or catastrophic droughts. The disappearance of the Mochica civilization could also be due to large-scale social disruptions, for example, due to a loss of confidence in the priest caste or to discredit of the ruling classes.

Fig. 4.12 Mochica ceramic representing a sailor riding a fish. Moche ceramics are easily recognizable because of the predominance of certain colors such as red and black on a cream background. Author's photograph taken at the Larco museum in Lima

Fig. 4.13 Mochica ceramics often depict scenes from everyday life, including sexual relations. Author's photograph taken at the Larco museum in Lima

4.8 The Reign of the Mountain States: Tiwanaku and Wari

'God with Scepters' and Cyclopean Statuary

Lake Titicaca is the highest navigable lake on the planet (Fig. 4.14). The surrounding plateau is dominated by mountains and represents the largest expanse of arable land in the Andes. It has been occupied by humans for millennia. Tiahuanaco in Spanish (or Tiwanaku in Aymara) is the highest ancient settlement in this region.

It was around the sixth century CE that two mountain states imposed their authority over a vast territory: these were the State of Wari (or Huari in Spanish) and that of Tiwanaku, with cyclopean architecture and an impressive statuary. Iconographic themes similar to those found at Tiwanaku were discovered at Wari and it seems that these motifs appeared simultaneously at both sites. This could imply that Tiwanaku was the capital of the region and Wari was a large prefecture controlling the north of the country.

Located near the shores of Lake Titicaca, Tiwanaku began to impose itself from 350 CE, extending its influence beyond the lake around 500 CE. The expansion of Tiwanaku was mainly directed towards the south. Initially, the

Fig. 4.14 Lake Titicaca. Author's photograph

Fig. 4.15 Typical altiplano farm. Author's photograph

power of this great metropolis was undoubtedly imposed more by its religious prestige than by its military authority. This state eventually encompassed the southern highlands, the coastal valleys of southern Peru and northern Chile, the Atacama Desert, and part of western Bolivia.

At the beginning, this large city on Lake Titicaca was dependent on agriculture, but, given the high altitudes (around 3800 m), it was only capable of producing a few cereals and a few varieties of potatoes (Fig. 4.15). Despite the use of a system of perfectly irrigated raised fields, this high-altitude agriculture was largely dependent on precise calendar planning of agricultural work to avoid the consequences of the periods of drought which regularly devastated the shores of the lake. The need to colonize additional lands in more temperate and warmer regions to exploit new agricultural, mining, and animal resources generated the expansionist thrust of the empire.

The three pillars of this expansion and development were the need for agricultural resources that could not be produced on the *altiplano*, such as corn, peppers, coca, and others that were adapted to these ecological zones, the optimal development of breeding (llama, alpaca), and the appropriation of mining resources (turquoise, lapis lazuli, etc.), all justifying the development of an effective network of communication routes to promote commercial exchanges between the different regions (the coast, *sierra*, *altiplano*, and the southern deserts). Administrative centers ensured the link between the provinces and the metropolis occupied by the ruling elite who planned and set up mechanisms ensuring complementarity between the various resources.

The subjugation of the different regions was also based on the imposition of a common religious belief, namely the cult of the 'god with scepters' of Chavín influence, which was carved on the Tiwanaku Sun Gate. Religious symbols, including this god, viewed from the front and in profile, are omnipresent in textiles and on ritual objects such as ceramics. The polychrome decorations of the terracottas also show winged, anthropomorphic, and zoomorphic characters, as well as representations of animals like snakes, pumas, and condors.

Stone sculpture was a specialty of the Tiwanaku culture, with immense monoliths found particularly around the temple of Kalasasaya (today in Bolivia). These large statues of dignitaries (the largest measures more than seven meters high) demonstrate considerable expertise on the part of their builders.

Tiwanaku's power over the southern Andes lasted a long time. While the expansionist movement of this state indeed began around 500 CE, the decline was only announced around the year 1000 CE. The reasons for its decline are not clear. The most likely hypothesis is a succession of serious droughts that would have hit agriculture hard and caused the populations of the highlands to move elsewhere.

The 'Militarism' of the Waris

The capital of the Wari empire, located 2800 m above sea level, was near the present-day city of Ayacucho in Peru. It controlled immense areas from the region of Cuzco to that of Cajamarca. This civilization was born in the sixth century CE and expanded towards the north and towards the religious center of Pachakamaq, which nevertheless retained significant autonomy. It was contemporary with the Tiwanaku culture. The two cultures had many points in common, particularly in the arts, and have only been differentiated quite recently by archaeologists. The Wari empire extended as far as Chancay on the coast and Huaylas in the highlands.

Wari was one of the most imposing cities to spring up in ancient Peru. This capital had an efficient sewage system and was served by an elaborate road network. It testifies to a high level of militarization, as can be seen from the fortifications, wall decorations, and funerary monuments. The Waris were accomplished builders, putting up many public buildings, temples, and residences. The settlements were protected by imposing surrounding walls and included administrative buildings, barracks, stores, residential areas for important people in the city, and spaces reserved for artisans. The architecture was imposing, meeting the military needs of an empire that clearly intended to ensure its defense.

The Waris developed a system of terrace cultivation in order to increase agricultural productivity in the mountainous areas.

Artistic production by the Waris and Tiwanaku was quite similar. Both used polychrome paintings, highlighted in black, depicting both human beings and deities. Polychrome jars featuring an anthropomorphic neck were characteristic of the Wari style. Another typical item was the *qeros*, used for libations based on *chicha*.

The Wari empire was ruled by chiefs assisted by an extensive bureaucracy. They dealt with religion and war, but also organized festivities. The iconography depicts a ruling elite wearing bright tunics and characteristic four-cornered hats, possibly associated with scepters, typical symbols of power. These tunics were doubtless produced in small numbers, since they required significant labor and a considerable amount of camelid wool, and they were probably only made for aristocrats or heads of clans or communities.

The Wari Empire collapsed between 800 and 1000 CE, probably following rivalries with the kingdom of Tiwanaku, which caused sporadic clashes. It could be that environmental factors contributed to this decline, but it is difficult to determine whether they were the predominant cause of the end of these Middle Horizon states. As the Tiwanaku and Wari declined, the world of the central Andes experienced a new period, governed by regional states.

4.9 Sicán and Chimú Civilizations

Sicán Culture and the Country of Naymlap

During the Late Intermediate Period (900–1450 CE), the fall of the great highland powers was followed by a period of instability. The highlands region and the coastal zone experienced an alternation of smaller political entities, and several regional kingdoms emerged. On the coasts in the far north of Peru, the Sicán or Lambayeque culture prevailed where the Mochicas had been. It flourished in the La Leche valley from 700 to 1300 CE. Batán Grande, which is located near the current city of Chiclayo, became the political and religious center of this culture. They were essentially a people of sailors and traders whose civilization developed north of the Chimús community and that of Chancay.

The excavations carried out around Batán Grande have discovered *qeros* in precious metal, gold masks contained in funerary *fardos*, pearls, emeralds, semi-precious stones, and shells in the tombs of local lords. Many objects were decorated with seabirds, fish, or fishing scenes. The mythical character of Naymlap is omnipresent in the decorations and appears in particular on the *tumis*, the symbolic sacrificial knives, often made of gold and set with precious stones.

Naymlap was the legendary founder of the dynasty of sovereigns of the Lambayeque valley, according to the chronicle of Miguel Cabello de Balboa, the author of *Miscelánea Antárctica, una historia del Perú antiguo* [1586] (1951) (see Sect. 3.3). Represented with the wings of a seabird, this legendary character appeared on the coasts of Lambayeque aboard a boat made of woven reeds. Coming from the Pacific, he was accompanied by his wife Ceterni, his harem, nobles, and servants. He also carried a green stone idol called Yampallec which represented him and which gave its name to the Lambayeque valley. He settled in Chot and became king after seizing the territory of the Lambayeque region. When he died, he flew away like a bird. One of his sons, Cium, succeeded him, and his other sons settled in the region.

Around the nineteenth century BCE, we see the development of bronze production in the Lambaye region, which henceforth replaced bone and stone for the manufacture of tools. This mastery of metal opened up new perspectives for producing agricultural implements, developing irrigation networks, clearing land, and also manufacturing weapons of war.

The pyramids present at several sites played an important role in the culture of Lambayeque because, according to local beliefs, they established a connection between the lord and the mountain gods. According to these same beliefs, the major weather phenomena which occurred reflected the anger of the gods and indicated that the ceremonies organized on the pyramids and in the ritual spaces where the priests officiated had not fulfilled their role and had proven incapable of protecting the population. This inability of the authorities to fulfill their role as intercessors with the divinities generated discontent among the people and encouraged them to burn these buildings down. The capital of this culture, namely Batán Grande, was itself burned down, and its destruction sounded the death knell for Sicán power.

The Chimús and the Kingdom of Chimor

Chanchán and His *Ciudadelas*

The Chimús were the inhabitants of the kingdom of Chimor, whose capital was Chan Chan (or Chanchán), a city with adobe buildings, located not far from the current city of Trujillo. This city was founded by the inhabitants of the coastal region of Moche in around 850–900 CE. The Chimú civilization benefited from the remnants of the Mochica culture remaining in the region, in the Moche and Chicama valleys. It was during the following centuries that the Chimús empire reached its peak, dominating vast areas, bounded in the north by the Sechura desert.

The strip of territory occupied by the Chimús was located on the northern coast of Peru between the Pacific and the western foothills of the Andes. It is a desert region, crossed by rivers flowing from the mountains to the ocean and creating fertile oases. The plains were suitable for intensive agriculture thanks to well-organized irrigation and different agricultural techniques such as dams, underground aqueduct networks, and stepwells to extract groundwater. Indeed, one of the major achievements of this culture was the rehabilitation and development of irrigation networks. The Chimús cultivated beans, cotton, sweet potatoes, and papayas. Fishing was also a major resource, as was the breeding of camelids (llama, alpaca).

The capital Chan Chan was one of the largest and richest towns in pre-Columbian America. It would have had more than 80 000 inhabitants. The expansion of the Chimú empire took place mainly in the fourteenth and fifteenth centuries and extended well beyond the area of influence that characterized the Mochica culture. It was especially the rich regions of the south (Casma, Pativilca, Huaura, etc.) which attracted Chimú expansionism. At its peak, this culture extended along the coastline for a distance of more than 1200 km as far as Lima. The territory was dotted with fortresses guarding the valleys where garrisons were sheltered. The most famous of these is undoubtedly Paramonga. The expansionist thrust of the last Chimú sovereigns towards the east quickly clashed with the imperialism of the Incas and they were in fact conquered by the latter under the reign of *Tupaq Yupanki* and integrated into the Inca empire.

A founding legend relating to the royalty of Chimor carried through until the colonial era, like the one which prevailed in the kingdom of Lambayeque. It tells the story of a dynasty of sovereigns whose founder was called Taycanamu and which ended with the submission of King Minchançaman to the Incas after the reign of nine (or eleven) monarchs.

Chimús ceramics, used for ritual offerings or intended for domestic use, were characterized by a black monochrome, sometimes decorated with brown. This color was obtained by firing the pottery at high temperature in a completely closed kiln (Fig. 4.16). The art of Chimú textiles focused on spinning cotton and alpaca wool, then decorating the fabrics with feathers and silver and gold discs (Fig. 4.17). Fabrics were often painted with natural vegetable dyes from walnut shells, animal dyes like cochineal, or minerals like clay.

The Chimús seem to have forced blacksmiths from Lambayeque to work for them in Chan Chan, a way of operating which would later be implemented by the Incas. The metal objects made by these people were finely sculpted. A range of different techniques were mastered, such as plating, gilding, filigree, and lost wax casting. These techniques allowed the manufacture of very diverse

Fig. 4.16 Work of a Chimú artisan showing a fisherman, his lobster trap, and his catch in the form of a large crustacean. Author's photograph taken at the Larco museum in Lima

objects including bracelets, containers, figurines, *tumis* or sacrificial knives, and jewelry. The materials used for these purposes were copper, gold, silver, and various alloys like bronze. The Chimús also traded in shellfish, used for ritual purposes, notably spondylus fished off the coast of Ecuador. They also made wonderful ceremonial costumes, decorated with earrings, feather headdresses, necklaces, and pectorals.

Chimú society was hierarchical, with a powerful elite in the capital Chan Chan, at the head of the main centers of power. The ruling class lived in extraordinary luxury, as attested by the abundance of ceramics from this period, along with gold jewelry and the very rich furniture found in their graves. At the next lower level of power were the authorities of the fortified cities (*ciudadelas*) which were subordinate to Chan Chan. The secondary centers of power were responsible for managing land and water resources and organizing the use of local labor.

The capital had ten areas that were undoubtedly intended for kings and nobility, surrounded by very high adobe walls, painted or decorated with friezes, which gave them the appearance of fortresses. The decorations represented geometric or zoomorphic figures. The organization of the *ciudadelas* suggests that they were the residences and probably the mausoleums of the kings of

Fig. 4.17 Chimú style fabric decorated with anthropomorphic figures and stylized felines. Author's photograph taken at the Larco museum in Lima

Chimor. Within the enclosures, fairly modest rooms (the *audiencia* or audience chambers) would have been the residences of the royal administrators. Most of the population occupied barrios outside the *ciudadelas*, which included housing with areas for cooking and craftwork and premises for storing goods.

In the different regions of the kingdom, local deities were worshiped in sanctuaries of varying importance. These temples contained sacred objects to which worship was devoted. Some of them were described by the Spanish chronicler Antonio de la Calancha [1638] (1981) (see Sect. 3.3).

Chimús Astronomy

The Chimús worshiped the Moon, not the Sun, which they considered too destructive, probably because of the merciless solar radiation permanently blazing down on the desert areas where they lived (Rowe, 1948). The Moon (called Shi) was a major deity. It was considered more powerful than the Sun because it appeared both at night and in the day, because it was known to influence the tides, and because, according to Chimú beliefs, it promoted the growth of plants. Moreover, its lunations were used to mark the passage of time. Lunar eclipses were interpreted as harmful and served as a pretext for expiatory rites. The great Temple of the Moon located in the Pacasmayo Valley was named Shi-an ('House of the Moon'). Human sacrifices took place there. In the hope that they would be deified after their death, children around five years old were sacrificed, and these rites were accompanied by offerings of colorful fabrics, fruit, and *chicha*. Animals were also offered as sacrifices to the Moon.

As the Sun (Jiang) was deemed inferior to the Moon, solar eclipses were interpreted as victories of the night star over the day star. The latter was nevertheless considered the father of sacred stones called *Alaec pong*. It was believed that these stones, undoubtedly natural rocks, came from the Sun and were the ancestors of the inhabitants of the region.

The sea (Ni) was also among the important deities invoked through offerings to benefit from abundant fishing or to prevent drowning. Whales, which impressed with their dimensions, were considered sacred animals.

The Chimús observed the stars and honored some of them. In Orion's belt (Pata), the two stars outside the group of three stars in a line, which are easily identifiable in this constellation, were interpreted as emissaries of the Moon, busy leading a thief (the central star) to the birds of prey represented by the four stars appearing at a lower level in the sky. The Pleiades star cluster (Fur), whose appearance coincides each year with the beginning of the harvest, was used to define the agrarian calendar so important for organizing work in the fields, and it was widely believed that this constellation watched over the harvests (Rowe, 1948).

4.10 Chancay and Chincha Cultures

After the collapse of the Wari civilization, Chancay culture developed from 1200 to 1500 CE in a region including four valleys of the central coast of Peru. These are the valleys of Chancay, Chillon, Rimac, and Huaura. These coastal areas have fertile valleys where agriculture can flourish. The center of this civilization was located about 80 km north of Lima. Political and religious activities were structured around centers like Chico and Pisquillo.

The Chancays are known for their ceramics, dotted with black geometric patterns on a white slip. The vases were obtained by molding, a technique inherited from the Chimús, and had a fairly massive shape. Goblets and jars were often decorated with simple geometric designs, but in some cases they had a more elaborate anthropomorphic, zoomorphic, or phytomorphic shape. We can also note the presence of female figurines on ceramics which had a funerary purpose and *chinas*, anthropomorphic jars with eyes outlined in black. Many examples have been found in the necropolises of Ancón.

The artisans of Chancay mastered the work of embroidery and refined the textile processing techniques developed by the Paracas culture. Their production was very colorful. Figurines (sometimes called 'dolls') made of fabric, gauze, and reeds were found in tombs, where they were placed near the mummies. The deceased were wrapped in richly decorated *fardos*. These figurines were intended as offerings: they recalled a loved one or a period in the life of the deceased (Fig. 4.18).

The Chinchas lived in the region near what is today Lima, and their culture flourished especially from the eleventh to the fifteenth century. Contemporaries of the Chimús, they spoke Quechua and interacted widely with this neighboring power. They thus exchanged sacred shells, fish, and dried algae for copper, gold, and even emeralds. These people used chinchillas (literally 'little chinchas') to make clothes, which later became luxury items.

The Chincha temples and palaces were built of adobe, the enclosures of the fortifications being covered with bas reliefs. Ruins of urban sites from this era

Fig. 4.18 In many Andean tombs, figurines (also called dolls) made of gauze or fabric have been found, placed near the deceased. Photo taken by the author at the Adolfo Bermudez Jenkins regional museum in Ica

such as La Centinela and Tambo de Mora, centers located on the Pacific coast, can be found along communication routes. Chincha ceramics are polychrome and include flared containers and jars with handles and long necks. The colors are often ochres, reds, and blacks. The iconography, which is also found on richly colored fabrics, combines geometric patterns and zoomorphism.

It was general Qhapaq Yupanki, the brother of the Inca *Pachakuteq*, who led the army which annexed the territory of the Chinchas in 1471, but the Chincha lord retained significant prestige even during Inca domination.

5

Birth, Evolution, and Splendor of the Inca Kingdom

5.1 The Inca Origin Myth

The Chroniclers and the Myths

When we consider the Inca myths, it is important to remember that none of the stories that have come down to us were written by indigenous people in their own language. The Incas did not use writing, and the texts which have reached us were written by Spanish chroniclers or Incas whose native language was Quechua or Aymara, but who spoke Spanish (Roza 2008). In addition to the fact that there were no written indigenous stories, it should be noted that the available narratives are based on archives taken from *khipus*, the bundles of knotted strings first used to preserve information about censuses or tributes, which were managed by the *khipukamayoqs*. The narrative information preserved on the *khipus* was transmitted during public ceremonies (see Sect. 3.4).

There were also bards who kept the genealogical chronicle and the stories of the exploits of the Inca kings and queens up to date, presenting this information to the court in a poetic manner whenever the king requested it. Both groups, the *khipukamayoqs* and the bards, served as informants for the Spaniards after the conquest, their various accounts being translated by a bilingual interpreter for preservation by a Spanish editor.

It is obvious that these stories are not characterized by unfailing scientific rigor and may have been amended in different ways following personal or more political motivations, certain natives wishing to promote their own lineage, or the Spanish conquerors wishing, in some cases, to devalue the Incas. It is therefore important to treat these stories critically (Urton 2004). A certain caution is also necessary with regard to the chronology of the stories, because all the texts date from after 1532, so this mythical corpus does not contain rigorous information about the chronology, even if attempts have been made, notably by the Spaniards and then by contemporary researchers, to determine the precise dates of certain events mentioned in the myths. Even the dates of the reigns of some kings in the pre-Hispanic era remain uncertain, and most chronologies rest on shaky foundations.

Among the Spanish chroniclers who are of particular interest where the Inca myths are concerned, we can cite Cieza de León, Diez de Betanzos, Polo de Ondegardo, Sarmiento de Gamboa, de Acosta, Cabello de Balboa, Garcilaso de la Vega, de Murúa, de la Calancha, Guamán Poma de Ayala, de Santa Cruz Pachakuteq Yamki Salcamaywa, and Cobo. More details about these individuals can be found in Sect. 3.3.

Wiraqocha and the Creative Triad

Most myths relate to the origin of the world center on Lake Titicaca. Several rather different accounts exist. Here we follow the narration of Cristóbal de Molina ('El cuzqueño'), who wrote the *Relación de las fábulas y ritos de los Incas* [ca. 1575] (de Molina 1989). The story begins when the world is already created. *Wiraqocha* is the creator god (Demarest 1981):

> Then came a deluge and floods that covered everything, even the highest mountains. The only survivors when the waters receded were a man and a woman who washed up in the Tiwanaku region. *Wiraqocha* ordered them to stay there as *mitimaes*.[1] The creator god then fashioned the different nations of *Tawantinsuyu* from the clay and painted them, decorating them with the clothes they were accustomed to wearing. He then created all the animals on Earth, male and female for each earth race. *Wiraqocha* dispersed the different creatures, keeping only two with him who were in fact his sons and whose names were *Imaymana Wiraqocha* and *Tocapo Wiraqocha*. These new humans were therefore sent from their place of origin, Lake Titicaca, to different places in *Tawantinsuyu*, staying first in the underworld to be reborn into existence in different places such as caves, springs, [...] *Wiraqocha* then ordered *Imaymana*, his eldest son who was close to him, to travel northwest from Lake Titicaca following a route via the forest and mountains. The youngest son *Tocapo Wiraqocha* adopted the coastal route while the creator god, *Qon Teqsi Wiraqocha*, moved northwest following the highland route. Crossing the different regions, the creative triad brought back

---

[1] Settlers transplanted by the Incas to a different region from the one where they originated.

the ancestors of the local peoples from caves and mountain peaks. *Wiraqocha* and his sons then reached the north-western confines of the empire towards the Ecuadorian coast and disappeared on the horizon.

According to some authors, the triad formed by *Wiraqocha* and his sons may arise from Spanish influences. This Andean triad would then be analogous to the Christian Trinity. However, this is only a guess.

Pachakuteq and the Cycles of Creation and Destruction of the World

The cataclysmic destruction of the world and its replacement by a new creation is a notion that features in Inca cosmogony, just as it does for other cultures, like the Aztecs in Mesoamerica. The notion of cycle is present in the term *pachakuteq* used by chroniclers to express the destruction of the inhabitants of the world and their replacement by a new race.

If we are to believe Guamán Poma de Ayala's version of world history in the seventeenth century [1615] (Guamán Poma de Ayala 1936, 1980), humanity experienced a succession of five ages or 'Suns,' each of these being characterized by a duration of a thousand years. This text reflects a mixture of Christian contributions, but also indigenous elements, such as the division of the territory according to the *ayllus* and the reference to the creator gods *Wiraqocha* and *Pachakamaq*.

The First Sun was the age of Wari-*Wiraqocha*-Runa or the cycle of men of *Wiraqocha*, characterized by an initial period of darkness. The Earth was deserted and stony, but *Pachakamaq* created light. *Wiraqocha* transformed the stones into men. People of this age only had access to very rudimentary technology and wore clothing made from organic materials (leaves, plants, etc.). Initially, all living things spoke the same language, with animals, plants, and humans living in harmony. But soon the situation deteriorated, the Sun disappeared, and the Earth plunged into darkness. There was a rebellion of objects and animals, and the language of plants and animals became incomprehensible to human beings. Thus ended the First Sun.

The Second Sun was that of the sacred men, the Wari-Runa. This period was characterized by a more advanced civilization than the first. People wore clothes made from animal skins. They lived peacefully, practising fairly rudimentary agriculture. They recognized *Wiraqocha* as the creator god. A prolonged eclipse terrified the people at the end of the cycle and this age ended in 'heavenly fire.'

With the Third Sun, civilization became much more complex. People were referred to as savages (Purun-Runa) and they wore clothes made from fabric.

In addition to agriculture, they also engaged in mining and jewelry-making. As the population increased, conflicts became more common. The people of that time honored the god *Pachakamaq*. The Third Sun ended with a flood.

During the Fourth Sun, the age of warriors, the world was divided into four parts. As the number of wars increased, people settled in the mountains, where they built fortified houses. The *ayllus* structure became common and living conditions improved considerably.

The Fifth Sun was the age of the Inca civilization. Guamán Poma de Ayala describes the main institutions of the Inca empire and its religious organization. The Incas began to revere supernatural spirits (or *waka willka*). This fifth period ended with the arrival of the Spanish.

The Mythical Origin of the Inca Empire

The history of the first rulers of the Inca people goes back, according to legend, to the origin of the world. Two founding myths explain the beginnings of the empire with quite different contents, although they nevertheless include certain shared features.

One legend is recounted by Sarmiento de Gamboa in the *Historia de los Incas* [1572] (Sarmiento de Gamboa 1942, 2007). He wrote his text for Francisco de Toledo, the fifth viceroy of Peru. According to this version, the starting point would be the wanderings of the four *Ayar* brothers (*Manqo, Kachi, Uchu,* and *Awqa*) with their respective wives (*Oqllo, Qora, Rawa,* and *Waku*) departing from Pacaritambo ('place of origin') in the valley of Cuzco, where *Ayar Manqo* founded the city which would become the seat of the empire. These figures claimed to be the sons of *Wiraqocha*, the creator god who had disappeared into the ocean once his creative work was completed, but who had announced that he would return to Earth.

A long time ago, there was a mountain in Pacaritambo called Tambo Toco (or 'house of windows'), in which there were three windows (or caves). The side windows were named Maras Toco and Sutic Toco and were the place of origin of two nations who later allied themselves with the Incas, namely the Maras and the Tambos. The central window was called Qhapaq Toco (or 'rich window') and it is from there that the Incas emerged. These three groups were born from the caves of Tambo Toco under the leadership of *Qon Teqsi Wiraqocha*.

The four brothers and four sisters who emerged from the main window of Tambo Toco were named *Ayar Manqo* (also called *Manqo Qhapaq*) and his sister/wife *Mama Oqllo*, *Ayar Awqa* and *Mama Waku*, *Ayar Kachi* and *Mama Qora*, and *Ayar Uchu* and *Mama Rawa*. The eight original ancestors then left with their entourage, going northwards towards Cuzco in search of fertile lands. They could identify such lands by planting in the ground a golden scepter which they had brought with them.

During the first step in their journey to Cuzco, *Ayar Manqo* and his sister/wife *Mama Oqllo* conceived a child. This boy was destined to succeed his father. He was called *Sinchi Roq'a* and was born near Cuzco when his ancestors had conquered the city.

When they reached Haysquisrro, a major event occurred regarding the group's future. *Ayar Kachi*, who had a reputation for being turbulent and cruel, disrupted the towns he passed through and disrupted the harmony of the group. It was thus decided to get rid of him using a trick. He was asked to search for objects left in the Tambo Toco cave, including a golden cup, seeds, and a miniature llama (*napa*), a symbol of nobility. At first, he refused but, following the authority shown by *Mama Waku*, he ended up obeying. *Ayar Kachi* was accompanied by an individual called Tambochacay ('Tambo, the entrance-barrer'). It was Tambochacay who imprisoned *Ayar Kachi* inside the cave using a huge boulder. *Ayar Kachi* was then transformed into a *waka*.

The ancestors then reached the surroundings of the Cuzco valley and stopped at a place called Quirimanta located at the foot of the Huanacauri mountain. Taking the golden scepter intended to test the ground and throwing it down from the top of the mountain, they found that it sank into the earth, indicating that they had reached the place they were seeking. When they prepared to descend into the valley, the youngest ancestor was transformed into a rock. This place later became a *waka*.

The remaining ancestors left Huanacauri for a place called Matao. At this place, *Mama Waku*, who was very skilled with the slingshot, hit a man from the town with a stone that she had thrown and killed him. She frightened the townspeople by showing them the victim's entrails. They fled and the ancestors entered Cuzco. They went to see the chief Alcaviça and explained to him that they had been sent by their father, the Sun. They were then able to settle in the city. *Manqo Qhapaq* took corn seeds that he had brought with him and sowed them in a field. It was in this way, by sowing the fields, that the Incas took possession of the valley.

When *Manqo Qhapaq* and his companions reached the square of Hunaypata in central Cuzco, *Ayar Awqa* was transformed into a stone pillar which later became a *waka*. *Manqo Qhapaq*, who is often considered the founding father of the Inca empire, gradually dominated the inhabitants of the surrounding villages, considered barbarians, and eventually brought them to a higher level of civilization.

This mythical story has the merit of giving a common origin to the four ethnic groups of the region who had decided to unite, namely the Sawasirays, the Allkawisas, the Maras, and the Incas.

According to another myth and the chroniclers de Murúa [ca.1615] (2008), Guamán Poma de Ayala [1615] (1936, 1980), and Garcilaso de la Vega [1609] (1945, 1969, 2000), *Manqo Qhapaq* was also the legendary founder of the Inca dynasty with its headquarters in Cuzco, but the approach followed was different. In this legend, the Sun god *Inti* created a man and a woman on an island in Lake Titicaca. Their names were *Manqo Qhapaq* and *Mama Oqllo* and they were given a golden scepter. So, according to this account the Incas originated from the *altiplano*, and in particular from Lake Titicaca. *Inti* sent this couple through the world of the ignorant and uneducated to bring civilization.

Being the son of the god *Inti*, *Manqo* himself was to be worshiped as the Sun god. At the place where his scepter was planted in the ground, *Manqo Qhapaq* and *Mama Oqllo* were to found an empire and bring together the scattered and poorly educated peoples. The couple headed north and the golden scepter was indeed stuck into the ground at Huanacauri. This was the place where *Manqo* founded Cuzco. Among the activities that the couple taught this new people were spinning wool and cultivating the land.

If we compare these two legends, it appears that the founder of the dynasty was in one case from the region of Cuzco and in the other from Lake Titicaca. On the other hand, we find the civilizing action of *Manqo Qhapaq* in both stories.

5.2 The True History of the Inca People

During the Recent Intermediate Period, after the decline of the Tiwanaku and Wari empires, the local ethnic groups in the Cuzco region engaged in incessant struggles to try to impose their supremacy and appropriate the most fertile territories. One such group was the Inca people, although we do not know whether they originated in the region or whether they came from elsewhere. Following conflicts with and enslavement of their neighbors, and thanks to intelligent alliances with other ethnic groups and aided by a high-level administrative organization, this people would eventually succeed not only in dominating the entire region, but also in establishing their supremacy over a territory including present-day Peru and also immense areas extending from the forested plains of southern Colombia to the temperate territories of central Chile (Fabre 2005).

At the beginning of the twelfth century, the Cuzco valley gradually became a center of the regional economy and their ceramics spread widely throughout the region. The valleys of the Cuzco region were favorable for cultivating corn due to the fertility of low-lying lands bordered by the Watanay and Tullumayu rivers, while cultivated areas could be extended by means of an efficient irrigation system.

Excavations of sites where we find the imperial architectural style suggest an origin that could date back to the fourteenth century. It is therefore probable that *Pachakuteq*, considered by Spanish chroniclers as the founder of a great Inca state, was in fact only the heir of an older tradition and a state which had already established its hegemony over the region (Métraux 1967; Pease 1995).

The beginnings of the Inca kingdom correspond to the reigns of the first eight sovereigns (Table 5.1). These do not present significant events and the details are somewhat blurred by the Spanish chronicles and by oral traditions which tend to glorify their origins. The first five sovereigns belonged to the *Hurin* dynasty, originating from the lower districts of Cuzco, and the three others came from the *Hanan* dynasty, which came from the upper part of the city. Cuzco was gradually enlarged and beautified, but the sovereign found it difficult to establish his authority over the region. It was when *Wiraqocha*, the eighth Inca king, ascended the throne that the situation changed considerably.

In fact, it seems that the Incas did not easily prevail against the Sawasiray, Allkawisa, and Maras ethnic groups. However, they gradually adopted their language, Quechua, as well as certain cultural characteristics of these peoples.

Table 5.1 The main Inca rulers according to Itier (2010) and Fabre (2005). Some dates may be approximate

King's name	Dates	Period
Manqo Qhapaq		Pre-imperial
Sinchi Roq'a Inca	ca. 1230–ca. 1260	
Lloq'e Yupanki Inca	ca. 1260–ca. 1290	
Mayta Qhapaq Inca	ca. 1290–ca. 1320	
Qhapaq Yupanki Inca	ca. 1320–ca. 1350	
Inca Roq'a	ca. 1350–ca. 1380	
Yawar Waqaq Inca	ca. 1380–ca. 1400	
Wiraqocha	ca. 1400–1438	
Pachakuteq Inca	1438–1471	Imperial
Tupaq Inca Yupanki	1471–1493	
Wayna Qhapaq	1493–1527	
Waskar Inca	1527–1532	
Atawallpa	1532–1533	
Manqo Inca	1533–1545	Spanish conquest
Sayri Thupaq Inca	1545–1558	
Titu Kusi Yupanki Inca	1558–1571	
Tupaq Amaru Inca	1571–1572	

During the thirteenth and fourteenth centuries, *Sinchi Roq'a*, the son of *Manqo Qhapaq*, *Lloq'e Yupanki*, *Mayta Qhapaq*, and *Qhapaq Yupanki* played their role as warlords within the Cuzquenian confederation and launched conquering raids on neighboring villages. Through their successes, they then strengthened the role of the Inca tribe within the confederation. On the death of *Qhapaq Yupanki*, *Inca Roq'a* took control of the confederation and can therefore be considered the first Inca sovereign. After the reign of his successor *Yawar Waqaq*, *Wiraqocha*, who took power around 1400, had to put an end to the desire for independence of certain tribes. He conquered part of the upper Urubamba, then the territory of the Canas and the Canchis, thus giving the Incas access to the high plateau. The authority of *Wiraqocha* was now imposed over an extended region around Cuzco (Fabre 2005).

According to an indigenous story also relayed by Spanish chronicles, after a series of local conflicts, *Wiraqocha* allied himself with the Canchis and tried to impose his hegemony on the territories around Lake Titicaca, but their powerful neighbors to the north, namely the Chankas and the Andahuaylas, resisted. The Chanka tribe had joined forces with neighboring tribes to form a powerful confederation dominating the central area and the south of the cordilleras. Around 1438, they entered into conflict with the Quechuas and directly threatened the Cuzco region.

At the end of his reign, the king *Wiraqocha*, grew afraid and took refuge, in the company of his son Urqu, designated as his successor, in the fortress of Calca. Another son, *Pachakuti Inca Yupanki*, decided to defend the city of Cuzco and oppose the invaders by allying with neighboring fiefdoms, the Canas and the Canchis. Against all odds and through judicious and appropriate tactical maneuvers, in the company of his warlords Apu Maya and Wikatiraw, he annihilated the opposing army in a place called Yawarpampa (the 'Plain of Blood') and killed its two leaders. This victory put an end to Chanka power and expansionism in the region.

As *Wiraqocha* refused to take back the power, he left his son, who had vanquished the Chankas, to be crowned *Sapa Inca* (the 'only' Inca), but not before consulting the Sun god to obtain his approval. *Pachakuti Inca Yupanki* took the name *Pachakuteq* ('He who transforms the Earth').

Following this war, *Pachakuteq* pacified the region. He then entrusted control of the territories he had conquered to his brother Kapa Yupanki and, from 1445 to 1450, he went to war on the high plateaus. While annexing the lake region of the *altiplano*, Kapa Yupanki decided to conquer the north where the Anqaras, the Wankas, and the Waylas lived, going as far as Cajamarca, which was more than 1000 km from Cuzco.

To continue his expansion towards more distant lands, the Incas had to confront the king of the Chimús, who dominated the northern coast of Peru, and the lord of Cajamarca who reigned over the northern regions. *Pachakuteq* nevertheless succeeded in defeating these two peoples, and was thus able to reign over extremely large territories. *Pachakuteq*'s son *Tupaq Yupanki*, then *Wayna Qhapaq* continued the conquests towards the north of Ecuador and the east of Bolivia. Towards the south, the advances of the Incas were only stopped by the Mapuche and Chiriguano peoples, who put up fierce resistance.

Pachakuteq is considered the true founder of the Inca empire. He was in fact not only a great conqueror, but also a brilliant administrator of the conquered territories. He set up an efficient administration and army, developed the necessary means of communication to transmit messages across such an immense territory, and set out to rebuild Cuzco according to a new plan.

Tupaq Yupanki took over and pursued his father's work. Most of his reign was spent expanding the boundaries of the Inca empire. With his armies, he entered into campaigns against the peoples of the southern coast, seizing in particular the valleys of Chincha and Cañete, along with the ceremonial center of *Pachakamaq*. Around 1480, he launched his armies towards the Chaco in the south-east, where he conquered Potosi and ventured as far as Tucuman (Argentina) and also towards the south, where he captured part of what is now Chile. He is said to have founded several cities, such as Cajamarca, and built the famous citadel of *Saqsaywaman* near Cuzco. He was assassinated around 1493. His son *Wayna Qhapaq* then took power.

Wayna Qhapaq achieved new territorial conquests. Around 1511, he launched his armies into the far north of the empire, as far as Quito and even beyond, to fight different peoples, including the Karas and their allies. However, it was a campaign that proved to be longer and more difficult than expected. A dozen years later, the Kara territory was completely pacified and the imperial armies reached the current borders of Ecuador and Colombia. Around 1527, *Wayna Qhapaq*, who settled in the north of the empire at Tumipampa (or Tumebamba in Spanish), was struck down by a mysterious illness (probably smallpox), imported by Europeans into the New World, and which would prove to be deadly, since it would cause several hundred thousand victims.

5.3 Pizarro and the Spanish Conquest

Upon the death of *Wayna Qhapaq*, the kingship fell to one of his sons, *Waskar*, but another of his sons, *Atawallpa*, rose up against him. The struggle for the succession became fierce among the Incas (Fig. 5.1). The direct consequence of this uprising was that the Inca nation descended into civil war. The partisans of *Waskar*, who had remained in the Cuzco region, were defeated by the armies of the north. The capital was pillaged and its inhabitants massacred.

Atawallpa was informed of the landing of the Spaniards, commanded by Francisco Pizarro, at Tumbes in northern Peru. On November 15, 1532, the troop commanded by Pizarro entered the town of Cajamarca. The next day, at the request of the Spanish emissaries, the Inca in full force and accompanied by numerous soldiers went to the city to meet Pizarro. He was taken prisoner in an ambush set up by the Spanish, who had been hiding in the buildings surrounding the town square. They seized the Inca leader and mercilessly massacred his men. Pizarro demanded a huge ransom in the form of gold from the Incas to free their leader, but did not keep his promise. In fact, he

Fig. 5.1 The Inca *Atawallpa*, the last emperor of the Inca empire. Detail of a painting exhibited at the National Museum of Archaeology, Anthropology, and History of Peru in Lima. Author's photograph

had him executed for fratricide because, in the meantime, *Atawallpa* had had his brother *Waskar* assassinated, having suspected him of collusion with the Spaniards. Pizarro then went to Cuzco to make an alliance with the generals of *Waskar*, and recognized *Manqo Inca*, a young son of *Wayna Qhapaq*, as lord of the Incas.

At the beginning of his reign, *Manqo Inca* pushed the forces of Quito towards the north and reconquered the city which remained under the domination of one of *Atawallpa*'s generals. The idea of the new Inca leader was to use the Spanish conquerors to reestablish his authority over the entire empire and end the war of succession. Following in particular the pillaging of the temples of Cuzco by the conquistadors, he quickly came into conflict with the latter. *Manqo Inca* then tried to seize the cities of Lima and Cuzco where Spanish garrisons were established, but he failed in both cases because Pizarro managed to mobilize the Wayllas, Wankas, and Kañaris in his favor during the assault on Lima, and the tribes of Chachapuyas and Kañaris during the siege of Cuzco. The failure of the assault towards Cuzco in 1537 effectively sounded the death knell for the Inca empire.

5.4 The Last Gasps of the Empire

Manqo Inca and his last allies took refuge north of Cuzco in a mountainous region, the Vilcabamba cordillera, in order to escape Spanish domination. The Inca elite then divided themselves between resistance to the invader and collaboration with the latter, hoping in this way to benefit from prebends and largesse which would allow them to maintain an enviable status. It did not take long for internal quarrels to break out between the conquerors and between them and the holders of royal power. In 1545, *Manqo Inca* was assassinated by a Spanish refugee in Vilcabamba. His son *Sayri Thupaq* succeeded him and the Spaniards granted him a vast domain in Cuzco. *Titu Kusi Yupanki* then took charge of the Vilcabamba resistance. Upon his death in 1571, power was passed to *Tupaq Amaru*, another son of *Manqo Inca*. In 1572, Viceroy Toledo launched a Spanish expedition against Vilcabamba, aided by the Kañaris and the Chachapuyas. The execution of *Tupaq Amaru* finally put an end to Vilcabamba's resistance.

6

Cosmogony at Tawantinsuyu and the Inca Pantheon

6.1 *Tawantinsuyu* or the World of the Four Districts

The Cuzco Valley and, by extension, the Inca empire were divided into four parts (the *suyus*). The city of Cuzco was at the junction of these four parts and constituted the center of the cosmological order of the Andean world. The four districts were distinguished by the main roads which left the city towards the different regions of the empire. The latter was called *Tawantinsuyu* (which in Quechua means 'the empire of the four directions') (Espinoza Soriano 1997; Pärssinen 2003). The names of the four districts were derived from the names of the peoples who inhabited them. Thus in the southwest we find *Kuntisuyu* ('region of the Kuntis,' a people inhabiting the maritime slopes of the Cordillera), in the northwest *Chinchaysuyu* ('region of the Chinchas,' a coastal state populated by the Chinchas), to the south *Qollasuyu* ('region of the Qollas,' a people who occupied the north of Lake Titicaca), and finally *Antisuyu* ('region of the Antis,' who occupied the region to the northeast of Cuzco).

The neighborhoods were grouped two by two: on the one hand *Hanansuya*, the upper part of the city which included *Antisuyu* and *Chinchaysuyu* and, on the other hand, *Hurinsaya*, the lower part which included *Qollasuyu* and *Kuntisuyu*. The separation between upper Cuzco (or *Hanan Qosqo*) and lower Cuzco (or *Urin Qosqo*) was not only spatial but also social, economic, and political. In the cosmic vision of the Andean world, *Hanan* was identified with what is masculine and *Urin* with what is feminine. *Urin* corresponded to the domain of priests and intellectuals, often dressed in black and wearing a Moon on their chest. They were responsible for the *Qorikancha* (or Coricancha), the Temple of the Sun. *Hanan* was the domain of the military and politicians. They were often dressed in red and wore a Sun on their chest. They controlled the *Sacsaywaman* site (Itier 2010).

The Spanish chronicler Cobo [1653] (1956) also mentions a more complex partition of the city using the system of *seq'es* (or *ceques*) to which we will return later (see Sect. 8.4).

The center of the city of Cuzco, located at approximately 3400 m above sea level, was concentrated at the foot of a hill between two rivers, the Watanay and the Tullumayu. The population of the city and its suburbs probably reached 100,000 inhabitants during the Inca era. In the center lived mainly the nobles and their servants. There were temples richly decorated with gold objects and palaces which contained the mummies of deceased sovereigns. It was a city with a marked religious character, where large amounts of offerings were made on a regular basis by pilgrims.

Housing consisted in rectangular buildings with one room and a thatched roof, arranged around a courtyard and surrounded by a wall to form an enclosure. The streets were narrow and lined with high walls built of carefully squared stone. The center of the city had two large contiguous squares where events such as ceremonial banquets and military parades were organized. At the top of a hill overlooking the city stood the enormous fortress of *Saqsaywaman*, built of stone and raw bricks and comprising numerous buildings, including a cylindrical tower, inside a fortified enclosure (see Sect. 7.2).

With a little imagination, the city of Cuzco had the shape of a puma, a sacred animal for the Incas. The hill and the *Saqsaywaman* complex formed its head. The body of the animal was bounded by the Tullumayu and Saphy rivers, and the tail corresponding to the confluence of these two rivers. The space between the animal's front and rear legs marked out the *Hawkaypata* ('crying place' or 'resting place'), which was used for ritual banquets, and the *Kusipata* ('place of festivities') where the military parades took place.

At the head of the state was the *Sapa Inca*, the absolute sovereign and the son of the Sun. Half the territory of Cuzco, called *Cuzco Hanan*, was ruled by the *Sapa Inca*, who held supreme political, economic, and military power. The management of the other half of the city was ensured by the priestly power, whose supreme authority only replaced the reigning Inca in exceptional circumstances. The political organization in Cuzco was therefore a sort of attenuated diarchy (Fabre 2005).

Each of the four regions of the empire was headed by a governor chosen by the *Sapa Inca* among the aristocrats. The sovereign was at the head of a *panaka* which was founded at the time of his accession to the throne and which included a group of nobles ensuring the main responsibilities and charges of the state.

The common people (*hatun runa*) were essentially peasants who lived in the heart of a small local community (*ayllu*). An *ayllu* or lineage was made up of the descendants of a founding ancestor. The mummified body of this ancestor was preserved to be regularly honored. The lineages, whose size varied but which could bring together as many as several hundred households, had above all an economic vocation, being made up of farmers, potters, fishermen, and so on. The members of the *ayllu* cultivated an estate (*marka*) belonging to the community, the relations of production and distribution of goods being regulated by principles of assistance and reciprocity (Fabre 2005).

The members of the *ayllu* offered part of their work, controlled by civil servants, to the sovereign, but in exchange, in the event of major problems (starvation, earthquakes, floods, loss of animals, etc.), they received help from the *Sapa Inca*, who would provide them with goods kept in state warehouses (the *qolqas*). Among a range of tasks that the *hatun runa* carried out in the service of the community, we may mention working the land, exploiting mines, manufacturing tools and utility objects, repairing state property (roads, buildings, bridges, etc.), and also military service, at least in certain regions (d'Altroy 1992). A man called the *kuraka* or *cacique*, whose appointment procedure could vary depending on the region (election, belonging to a privileged family, etc.), exercised his authority over the whole of the given *ayllu*. His main role consisted in organizing work and distributing tasks within the lineage.

Based on reciprocity, the collective labor system included three kinds of service: the *ayni*, which corresponded to the services provided within the community (sowing, harvesting, etc.), the *mink'a*, which consisted of collective work essential to the well-being of all (digging canals, laying out terraces, building temples, etc.), and the *mit'a*, which brought together the activities planned by the imperial authority (road repair, bridge maintenance, work in mines, etc.). State officials, the *khipukamayoqs* (*khipu* masters), played the role of accountants and had to report periodically to the *Sapa Inca* on the economic situation in the various parts of the empire. Inca society included structures larger than the *ayllu*. These larger entities (*llaqta*) were themselves grouped into larger structures which had centralized authority and which the Incas made into provinces.

Merchants and artisans also played a considerable role in Inca society. Following the contacts they had with distant peoples, they sometimes served as ambassadors or informants. Some of them, among the most talented or the most experienced, were automatically transferred to the capital Cuzco where, as a contribution to the empire, they produced prestigious goods, which guaranteed them a relatively privileged status (Figs. 6.1 and 6.2).

Fig. 6.1 This finely engraved gold plaque attests to the metalworking skills of Andean artisans. Author's photograph taken at the Larco museum in Lima

Fig. 6.2 Finely crafted necklace made from gold and precious stones. Author's photograph taken at the Larco museum in Lima

Yanas were generally prisoners of war who were assigned to domestic chores and agricultural work. They had to serve their master all their lives, having lost the right to belong to a community. However, they were allowed to own agricultural land and movable property.

Andean agriculture was intensive and made the most of the country's mountainous topography. Given the limited availability of arable land, farmers developed terraced crops (*andenes*). These were irrigated using techniques that made

Fig. 6.3 Woolwork was a major occupation of the 'Virgins of the Sun' during the Inca era. It still occupies many women in contemporary society. Author's photograph

the best use of the mountain slopes. They also cultivated small raised fields (*camellones* or *waru waru*), separated by canals where irrigation water flowed, a technique practiced in the *altiplano* and *puna* (Itier 2010).

The manufacture of fabrics was an important activity. Quality fabrics, intended for the elite, were made from vicuña wool. This work was entrusted to the *akllas*, or 'Virgins of the Sun,' who were confined in 'convents' and whose role was to serve as secondary wives or as 'rewards' for services rendered by members of the ruling class (Fig. 6.3).

6.2 Religious Practice and Beliefs of the Incas

The information we have about the religious beliefs of the Incas comes mainly from Spanish and Peruvian chroniclers of the sixteenth and seventeenth centuries.

A *panaka* was a lineage, a family group composed of all the male descendants of an emperor (Hernández Astete 2008). The Incas applied the principle according to which the property of a dead person remained with him after his disappearance. The *panaka* was responsible for maintaining these, and this maintenance was necessary to provide a means of subsistence for its members. This explains the cult devoted to mummies, which were kept in a palace and placed on a throne. These mummies were richly dressed and had servants who took great care of them.

In ancient Peru, the bodies of the deceased were dried and kept in a place protected from humidity. The dehydrated body was not necessarily embalmed but, subject to the dryness of the Andean climate, it could sometimes be preserved for several centuries. Many mummies were confiscated by the Spanish in order to bury them as part of their fight against 'idolatry' (Bernand 2010). The founding ancestor of the *ayllu* was particularly venerated. The mummy was periodically taken out of its tomb to make it drink, eat, and participate in the festivities of the living (Fig. 6.4). In particular, large gold, silver, or terracotta vases were placed in front of the mummy and filled with corn beer (*chicha*) to show the deceased that he could participate in the libations. The ancestor was also asked important questions during annual festivals that the *ayllu* dedicated to its founder. Through this cult, the descendants of this highly honored

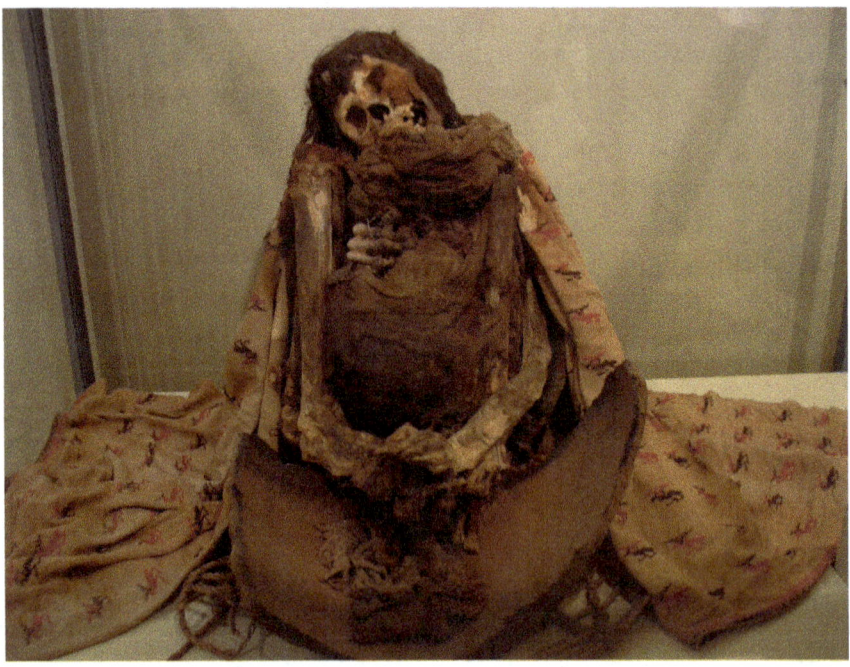

Fig. 6.4 Mummy from the Adolfo Bermudez Jenkins Regional Museum of Ica. In the Inca era, mummies were periodically extracted from their tombs to participate in the festivities of the living. Photograph taken by the author

6 Cosmogony at Tawantinsuyu and the Inca Pantheon

individual asserted their rights to the lands of the *ayllu*, but, in exchange, the ancestor had to ensure the fertility of the cultivated lands.

The Inca practice of passing royal property to family members other than the crown prince encouraged the latter to increase his property and established the limits of this system, because too much property belonged to the dead. It was *Waskar*, the great-grandson of *Pachakuteq*, who decided to abolish this custom. However, this generated much discontent and contributed to destabilizing the empire.

The ancestors of the lineages came from a particular place (source, cave, mountain, etc.) called the *paqarina* or 'place of appearance,' which was itself considered a sacred place and made the object of worship. One basic concept in the pre-Hispanic Andean belief system concerns the relationship between the sacred and Nature. Water, the sky, plants, and animals were all imbued with sacredness. Anything that had religious attributes and was venerated was called a *waka* or 'sacred thing.' These places where the supernatural manifested itself might be a wood, a rock, a spring, a temple, the statue of a god, or even a mountain.

The higher deities were known to project part of their being onto men, herds, and cultivated land, thus giving them part of their vital force, energy, and fertility. They then became *kamaq* ('who carries on his essence'). Thus the Sun was reputed to be the *kamaq* of the Incas. To take full advantage of the benefits emanating from the *kamaqs*, these powers had to be honored. This is why the shepherds worshiped and made offerings to the Llama constellation in order to obtain dynamic and fertile herds. Each individual benefited from positive flows emanating from various deities, each *kamaq* communicating to its receiver qualities that were specific to him and through which he differentiated himself from other people.

The *Sapa Inca* was a sacred figure. He was the son of the Sun, and as such he was worshiped during his life, but also after his death. He appeared as the great provider of vital forces to the deities in order to obtain from them the energy necessary to get through difficult circumstances. To this end, religious leaders carried out bloody sacrifices, particularly of animals but also of humans, in the hope of warding off calamities and catastrophes. Among the offerings, we find, in addition to animals (llamas, alpacas, etc.), agricultural products (corn, coca), pieces of goldwork (jewelry, decorations), valuable stones or precious textiles, shells, and so on, which were sacrificed during ceremonies taking place at different times of the year.

6.3 Inca Cosmogony

Inca cosmogony was based on the conception of a universe divided into three parts, three superimposed worlds between which priests ensured the communication. The upper world (*Hanan Pasha*) corresponded to the sky, the place where the deities were concentrated. It was symbolized by the condor, a mythical bird which soars high in the sky. The lower world (*Urin Pacha* or *Ukhu Pasha*) was the domain of the dead, materialized by a serpent. Finally, *Kay Pacha* was our own universe, the world where we live, symbolically represented by a puma. These different places, the seat of uncontrollable natural powers, communicated by rainbows and lightning. Reptiles represented both the upper world, the Milky Way being likened to a snake, and the lower world, snakes seeming to come from the ground or emerge from the waters. Lightning, which established the connection between the celestial world and the terrestrial world, was also materialized by an ophidian.

The Andean conceptions of the world of the Inca period have not completely disappeared and find an extension in the neo-Andean world view currently popularized by shamans. *Pachakamaq* corresponds to primordial energy, the energy at the origin of the cosmos and Nature. This power is unrepresentable and its name is pronounced with respect and devotion. Its manifestations appear indirectly, the most visible and immediate being the Sun (*Inti*), which would not exist without primordial energy.

Pachamama (or *Mama Pacha*) represents Mother Earth, the source of fecundity and fertility, but is also the seat of very violent and cataclysmic disturbances which occasionally manifest themselves through earthquakes, floods, and volcanic eruptions, for example. This goddess could be benevolent when satisfied or vindictive when she had not received the tributes she considered to be her due.

Andean spirituality imposed absolute respect for Mother Earth, whose violence could be extreme and whose power manifested itself in a particularly spectacular way in the different regions of Peru, whether in the Andes Cordillera, in the fertile coastal valleys located in the middle of the desert, or the lush vegetation of the Amazon rainforest. The considerable energy stored in Nature demanded the utmost respect, and the remarkable power of the cosmic forces encouraged humility.

In the context of Inca cosmogony, natural powers were considered as active elements. This was so not only as regards the Sun, the Moon, the constellations, thunder, and volcanoes, but even when it came to stones, rocks, lakes, and the like. Caves, caverns, and underground galleries gave the living access to the world below. The Incas attached particular importance to many rocks or stony places because they considered that the stones radiated and therefore became vectors of telluric energy.

The world continually evolved in a cyclic way, with the cycles usually ending in cataclysmic events. The invisible universe evolved permanently towards the visible world with the establishment of a balance between antagonistic forces. This balance was found in different aspects of Nature, such as the Sun and the Moon, the father and the mother, the masculine and the feminine, for instance, and it had to be preserved at all cost.

6.4 The Complexity of the Inca Pantheon

During the expansion of their empire, the Incas did not force the conquered peoples to abandon their beliefs and their divinities. The different cultures integrated into the kingdom saw their gods combined with those of their rulers. The deities were associated with the natural and cosmic environment in which these people lived. The religion of the Incas was in fact a world view incorporating the regional variations of different peoples (Marín-Dale 2016). The god *Wiraqocha* played a particular role as master of the elements, as he already appeared in the civilizations of Wari and Tiwanaku. Inca religion incorporated the heritage of previous cultures in the same way that the kingdom appropriated the artisanal know-how of conquered regions. This religion was polytheistic with a pantheon comprising many deities, but it was also pantheistic, establishing a connection between human beings, the phenomena and manifestations of Nature, and the gods. The importance given to natural elements encouraged people not to disrupt the order of things (Conrad & Demarest 1984; Scott 2005).

Religious beliefs could not be dissociated from the smallest gestures of everyday life. The gods lived in the sky and on Earth, and each had specific functions that justified its interventions in different aspects of daily life. Many deities appeared in the form of inanimate objects or natural elements, such as mountains, rivers, caves, lightning, the Sun, the Moon, etc. Some deities appeared in the form of animals (birds, felids, snakes) and were likely to adopt human behaviors or even feel human feelings (Métraux 1967; Steele & Allen 2004).

Pachakamaq and the Main Deities

A major divinity of the Cuzquenian universe was *Wiraqocha* (the 'Lake of Daybreak') (see Sect. 5.1). *Wiraqocha* was already revered by cultures before the Incas, namely those of Chavín and especially Wari and Tiwanaku. He was the primordial and universal god of the Incas, the creator god of lightning and storms. *Wiraqocha* presided over agriculture and the irrigation procedures associated with it. This divinity was directly linked to the springs and lakes which supplied water to the irrigation canals. It was he who created the sky, the Earth, the Sun, the Moon, and the first human race, which had its origins around Lake Titicaca. He is often compared to the other archaic divinity, *Pachakamaq*, in the founding myths. *Wiraqocha* was the main god of Inca mythology, but the emperor *Pachakuteq* designated *Inti*, the Sun god, as the supreme deity of the local pantheon.

Pachakamaq (the 'creator of the world') was the creator god of the peoples established along the coasts of Peru. He was the master of earthquakes and also the deity of the Earth in general, and was sometimes identified as the consort of the *Pachamama*. Subsequently, he was identified with *Wiraqocha*, the supreme god of the Incas. *Pachakamaq* is also the name of a major archaeological site located about thirty kilometers from Lima, where remains belonging to different civilizations have been unearthed from the time of the Lima culture (200–600 CE) until the Incas. There is a temple dedicated to *Pachakamaq* (the Painted Temple) and another dedicated to the Sun (Pozzi-Escot et al. 2017) (see Sect. 7.5).

Pachamama (also called Mama Pacha) was the primordial Earth Mother and was associated with fertility and agriculture. She presided over sowing, planting, and harvesting and was also believed to be responsible for earthquakes.

Another supreme divinity was the lightning god, who could not be dissociated from lightning and thunder. He was a god associated with atmospheric phenomena and rain formed from the waters of the upper world, which flowed through the sacred river that constituted the Milky Way. This god was called *Illapa* ('lightning') in Cuzco and *Lliwyaq* ('clear sky') in central Peru. He was invoked as protection against drought. *Wiraqocha* turned out to be more the protector of the inhabitants of the valleys cultivating corn, for whom irrigation was important, while *Illapa* was honored primarily by the inhabitants of the *puna*, who were mainly involved in the cultivation of potatoes and the practice of livestock breeding. At high altitudes, storms accompanied by lightning are in fact more frequent than in the plains. *Illapa* was also invoked during battles, to enlist his power and ardor. He was depicted as a warrior in the sky, wearing shining clothing and carrying a sling. The lightning was supposedly caused by his clothing and the thunder by the sound of his slingshot.

Inti was the Sun god. He was the son of *Wiraqocha* and every day he traveled the sky from east to west to dive into the sea and be reborn the next day. Dispenser of heat and light, he was the protector of the Inca people and was represented as a disc with a human face surrounded by rays. *Inti* was considered a very important deity in the Inca pantheon and his cult was extended throughout the empire. The Inca emperors also claimed their descent. He was represented in the *Qorikancha* of Cuzco in the form of a golden statue in human form, the *P'unchaw* ('The Day'), surmounted by a disc also made of gold. He appeared in the form of a seated child with a band around his head, ornaments on his ears, and wearing a pectoral. Rays emanated from his back.

In the Inca cosmovision or world view, *Inti* could take three different forms, namely the father, *Apu Inti* or the Sun-Lord, the son, *Churi Inti* or the light of day, and the brother, *Awqi Inti*, who also figured at the center of the cult of the Sun in relation to the founding of the empire. The first two were associated with the solstices and were celebrated during imposing ceremonies: *Inti Raymi*, during the winter solstice, materialized the worship of the Sun, while *Qhapaq Raymi*, during the summer solstice, corresponded more to a cult paid to the *Sapa Inca*.

Inti was highly revered and the most prestigious temples in the empire were dedicated to him. *Willaq Umu*, the high priest of the Temple of the Sun, the *Qorikancha*, located in Cuzco, was one of the most powerful people in the empire. With his wife and sister, the divinity of the Moon, *Inti* is said to have fathered *Manqo Qhapaq* and *Mama Oqllo*, the founders of the Inca dynasty. On this basis, the imperial power claimed its legitimacy from its direct filiation with the Sun. It was from *Pachakuteq*, the ninth Inca of the dynasty, that the cult of the Sun spread throughout the empire.

Inti's wife and sister was the Moon, or *Mama Killa* ('Mother Moon'), who was worshiped in the temple on the island of Koati, on Lake Titicaca, and celebrated in particular during the feast of *Qoya Raymi*. In Cuzco, a group of priestesses called *Aklla Wasi* (the chosen women) were in charge of her cult. *Mama Killa* was also the goddess of marriage and protector of women. She was considered to occupy third place in the Inca pantheon after *Inti* and *Illapa*, and was represented by a silver disk with the face of a woman. This was important in relation to the counting of time because different rituals were based on the lunar months of the Inca calendar. The Incas feared lunar eclipses because they believed that, during such an event, *Mama Killa* was being attacked by a wild animal (see Sect. 10.18).

In the Cusqueñian universe, the powers of the emperor's main wife (*Qoya*) emanated from the Moon (*Killa*) and those of the *caciques* came from the Morning Star (*Ch'aska Qoyllur* or 'the ruffled star').

Secondary Deities and Natural Phenomena

Among the rather numerous secondary deities in the Inca pantheon, most had a link with the stars or natural phenomena. They were regularly invoked to obtain good harvests or to ensure protection during disastrous climatic phenomena. In relation to the manifestations of natural phenomena, we may mention *K'uychi*, the Inca god of the rainbow, who was honored in the *Qorikancha* in Cuzco. The god of the rainbow embodied beauty and therefore became the god of nobility. There was also *Kon*, the god of rain and the south wind, who was one of the sons of *Inti* (the Sun god) and *Mama Killa* (the Mother Moon). Then there was *Pariaqaqa*, the god of water, torrential rains, and wildlife. The latter already featured in pre-Inca mythology, but he was later adopted and worshiped by the Incas. We may also mention *Apu*, the god of the mountains, to whom sacrifices were offered.

Very specific deities were associated with certain professions. This was the case of *Mama Qocha* ('Mother of the Sea'), who was the goddess of the sea and fish, as well as the protector of sailors and fishermen. She was important in the territories along the coasts, but in some parts of the empire, she was also considered the goddess of lakes and rivers. This was the case for Urkuchillay, too, a deity honored by Andean shepherds.

In the Andean environment, obtaining sufficient harvests was vital for the population. It is not therefore surprising that the Incas invoked several different goddesses for the protection of specific crops. Thus, *Acsumama* was the goddess of the potato. She was considered to be the daughter of *Pachamama*. *Saramama* was the goddess of corn, a plant abundantly cultivated in the Andes, and *Kukamama*, the goddess of coca. The pantheon also included *Kinwamama*, the goddess of quinoa, a familiar plant in the Andean world.

Other secondary deities were:

- *Cavillace* was a virgin goddess who ate a fruit that was actually the sperm of Coniraya, the Moon god. She became pregnant and gave birth to a son who was later transformed, along with herself, into rocks on the coast of Peru.
- *Ch'aska* or *Ch'aska Qoyllur* (Venus) was the goddess of dawn and dusk. She was also the protector of young girls.
- *Copacati* was the goddess of lake waters.
- *Eqeqo*, the god of abundance and prosperity among the peoples of the Peruvian and Bolivian *altiplano*, was the subject of sustained worship. Offerings were made to him by the Incas, notably dolls, to bring his blessing on the household. *Eqeqo* is a variation of *Ekhako*, the god of fortune in *Qollasuyu*.

According to legend, *Eqeqo* kept the Milky Way in a jug, emptying it to make rain.
- *Kilya* was the goddess of marriage.
- *Mallku* ('spirit of the mountain') was a deity who represented the spirit and strength of the mountain. He took the form of a condor.
- *Mama Allpa* ('Mother of the Earth') was the goddess of fertility. In iconography, she was depicted with multiple breasts.
- *Mama Nina* ('Mother of Fire') embodied the divinity of light, fire, and volcanoes.
- *Mama Wayra* ('Mother of the Wind') was the goddess of air and winds, protector of birds. She had a purifying role.
- *Paricia* was the god of floods. This may actually be another name for the god Pachakamaq.
- *Piguerao* (also called *Pikiru*) was the god of the night and demons.
- *Qoyllur* was the goddess of the stars.
- *Supay* was the god of death and the master of *Ukhu Pasha*, the underworld. With Spanish colonization, *Supay* was compared to the devil and was even invoked in this capacity by the natives.
- *Tulumanya* (also called *Turumanyay*) embodied the rainbow of the ancients, the first rainbow.
- *Urcaguary* was the god of metals, gems, and other precious elements from underground.
- *Yanañamca and Tutañamca* were gods of darkness and night. They ruled the world at its origin, before the gods took care of the Earth.

Mythical Creatures and Anatomical Implausibilities

In the Mesoamerican world, certain animals were considered mediators between the divine and the terrestrial worlds. This explains why those who held religious powers sometimes took the form of an animal. These concepts also seem to prevail in the Andean world, as evidenced by certain Mochica and Chimú ceramics, which often present anthropo-zoomorphic forms. They sometimes embody shamans in a zoomorphic aspect when they enter trances and come into contact with the divine world. The ceramics also show a wide variety of animals such as felines, deer, owls, and foxes, not in an entirely natural form, but with human characteristics. The deities of agriculture and fertility are sometimes depicted in the form of hybrid beings, mixing human characteristics with a representation of plants such as squash, beans, and corn.

The Lanzón of Chavín de Huántar (see Sect. 4.2) in the Northern Sierra of Peru is an enormous sculpted monolith which represents a standing figure with anthropomorphic features, but also with zoomorphic elements taken from felines and reptiles. The iconographic translation of the beliefs of the Chavín civilization generated fantastic beings to embody the divine. In order to reveal this divine world in an impressive and extraordinary way, the craftsmen created artistic combinations mixing the human being and the anatomical characteristics of animals such as the jaguar, the snake, the falcon, the condor, or the caiman, thereby creating anatomically implausible creatures.

The access portico to the Castillo is framed by two stone columns, each containing a sculpture representing a mythical winged human-looking figure combined with animal anatomical elements. On the columns, an engraved slab has motifs in the shape of mythical birds of prey, also with feline and reptile features.

On the Tello obelisk, we see the ancient 'jaguar man,' which presents a terrifying appearance. The image is complex and composite because the feline, the bird of prey, the caiman, and the man are mixed with plant elements and spondyl shells. These motifs, representing mythical beings, evoke the interweaving of the world of humans with that of animals and plants. The presence of caimans implies the interaction of the Chavín culture, in one way or another, with very distant Amazonian civilizations. It may be that an oracular cult was practised in Chavín whose reputation reached those distant lands.

The Raimondi monolith, in the same culture, is a stele representing a mythical character, holding a sparkling scepter in each hand. In a standing position, this hybrid creature has feet and hands with sharp claws, a jaguar's head, and snakes for hair and belt. We do not know the exact meaning of the images conveyed by this stele, and in particular the snakes which compose it.

The iconography of the ceremonial center in Chavín de Huántar did not have a purely decorative vocation. These anthropo-zoomorphic compositions, which intertwined the real and the unreal, also had the aim of revealing to the people the universe of the divine by representing fantastic creatures that would leave their mark on people's minds. The hybrid god, half man and half animal, with a terrible expression and flaming staffs, remained present in the history of the Andean peoples many centuries after this ceremonial site was abandoned.

There is also a mythical creature on the Sun Gate in Tiwanaku that may suggest a distant influence across time and space of the priests of Chavín, whose religious ideology could have carried this far. The architrave of the Sun Gate is decorated with a group of winged demons converging towards a hybrid central figure with hair composed of serpents and holding an anguiform scepter in his hand. This figure is often called 'the god of scepters.'

Chavín's influence can also be seen in the murals of Huaca de la Luna, a ceremonial site of the Mochica civilization. The hybrid and terrible aspect of the deities of the Chavín universe, modified by later civilizations, is assumed to embody the uncontrollable forces of Nature. It could be that these compositions recall the terrible disasters and droughts associated with the *El Niño* phenomenon. The priests would have widely popularized the worship of these deities from the Ancient Horizon and spread the idea that human sacrifices were necessary to appease their anger and fend off the merciless outbursts of violence they attributed to them.

7

Proximity of the Gods and Sacred Places

7.1 Cuzco, the Flagship City of *Tawantinsuyu*

Cuzco was the city of light of the Inca empire. The center of the Plaza de Armas is located at a latitude of 13° 29′ 1″ south and a longitude of 7° 58′ 45″ west, at an altitude of 3350 m. It is in the heart of this city that were once located the temples and palaces of *Tawantinsuyu*, an empire which at the height of its glory covered 1.7 million square kilometers.

Upon arrival in the city, the visitor cannot help but be struck by the many mountains that surround it. In fact, to the north, we find the hills of *Saqsaywaman*, Pukamoqo, T'oqokachi, Fortaleza, and Senqa, and to the northeast, we can make out Socorropata, Qorao, Pantorani, Picol, and Pachatusan. On the eastern side are the hills of Panpanusaka, Kunturqhata, and Saqsapata, while on the south side stand Kondorama, Araway, Choqo, Kachona, and Cheqollo, and to the west, finally, the hills of Pikchu and Apuyawira.

The hydrographic system which crosses the city is made up of many streams (Tullumayo, Saphi, Ch'unchulmayo, Wankaro, and more). They join to form the Watanay, which crosses the entire Cuzco valley to flow into the Vilcanota, also called Willcamayu or 'Sacred River' by the Incas, after receiving the waters of several other minor tributaries.

Today, what still fascinates visitors discovering this city are the remains of the marvelous Inca architecture (Bauer 2004). With their perfectly squared and adjusted stones, the walls attest to the mastery of the craftsmen, who only had rudimentary tools to carry out their work (Figs. 7.1 and 7.2). As one moves from the Plaza de Armas to the ruins of the fortress of *Saqsaywaman*, the city reveals itself as a sea of red roofs spreading out in a deep valley. The city today

Fig. 7.1 In Cuzco, the remains of the walls with their perfectly fitted stones attest to the mastery of the artisans of the Inca empire. Author's photograph

has around 300 000 inhabitants, the majority of the population being mestizos and Quechua-speaking indigenous peoples (Fig. 7.3).

If legend is to be believed, the city was founded around 1200 CE by *Manqo Qhapac*, who called it Cuzco (or Qosqo, the center of the world). Initially, this sovereign divided the city into four sectors, namely Q'ente *Kancha*, Yaramuy *Kancha*, Sayri *Kancha*, and Chunpi *Kancha*, to create a mythical structure of conquest and socio-economic and ideological domination of the region which later continued in the capital of the great empire *Tawantinsuyu*.

Around the middle of the fifteenth century, under the leadership of the *Sapa Inca Pachakuteq*, to whom a large part of the expansion of the Inca empire is attributed, the city received a new face. The Chancas had wanted to seize the city and it had been largely destroyed, but *Pachakuteq* had succeeded in repelling them. He decided to make it a majestic city, and Cuzco henceforth symbolized the glory and power of the kingdom with the construction of an ingenious water supply system, the erection of imposing buildings, and the irrigation of the fertile valley in which it was built.

This sovereign also created, organized, and restructured various social and religious institutions including the *mit'a*, the official cult of the Sun, and the

Fig. 7.2 Detail of a wall in Hutunrumiyoc Street in Cuzco. Author's photograph

functional urban model, among others. Regarding its urban structure, the city of Cuzco was divided geographically into two parts: an upper part and a lower part. Each of these sectors was in its turn divided into two sub-sectors: a right part or Allawka and a left part or Icho. The four *suyus* or parts of the city were bounded by the royal roads, thus reflecting, on a small scale, the division of *Tawantinsuyu*. The separation between upper Cuzco and lower Cuzco was not only spatial but also social, economic, and political.

In accordance with this duality, the city of Cuzco had two main squares: *Hawkaypata* in *Hanan Cuzco* and Rimaqpanpa in *Urin Cuzco*. There were also two Sun Temples: the *Qorikancha* in *Urin Cuzco* and *Saqsaywaman* in *Hanan Cuzco*, each with its own square and *usnu*. In addition, the city of Cuzco was a 'sacred space' comprising 358 *wakas* or sanctuaries of various types with religious and astronomical functions, arranged in a circle of radius more than twenty kilometers around the *Qorikancha*. The four main roads departed from *Hawkaypata* towards the four regions of *Tawantinsuyu*. If we are to believe the Spanish chroniclers, the square was surrounded by palaces and covered with a layer of earth which came from different regions of the empire and which had been brought there by travelers as a form of submission to the emperor.

Fig. 7.3 The city as it can be seen today from the ruins of the *Saqsaywaman* fortress. Author's photograph

Not far from the Plaza de Armas is the Convento de Santo Domingo, which was built on the ruins of the marvelous Inca temple, the *Qorikancha* dedicated to *Inti*, the Sun god, and which will be discussed later. After its pillage by the conquistadors, the ruins of the temple were offered by Juan Pizarro to the Dominicans, who built a church, a building quickly destroyed by the earthquake of 1650 but subsequently rebuilt.

At the time of the Incas, the central part of the city was an administrative and ceremonial district reserved for the imperial elite, places of worship, and the accomplishment of political tasks. Here were the large royal palaces. Thus in the upper part of the city, there was the palace of *Manqo Qhapaq*, and those of *Pachakuteq, Inca Roq'a, Waskar, Qhapaq Yupanki,* and *Wiraqocha Inca*. In the lower part of the town stood the homes of *Inca Wayna Qhapaq, Tupaq Yupanki, Inca Roq'a, Amaru Inca Yupanki, Mayta Qhapaq, Lloq'e Yupanki,* and *Sinchi Roq'a*.

The suburbs were inhabited by more ordinary people and they housed the agricultural populations who ensured the supply of the city with consumable goods. Thus the outskirts of the city were divided into thirteen districts with many inhabitants: Qolqanpata, Qantupata, Pumakurku, Munaysenqa,

T'oqokachi, Rimaqpanpa, Pumaqchupan, Kayaokachi, Chakillchaka, Pikchu, K'illipata, Karmenqa, and Wakapunku.

The predominant architectural model in Cuzco and in the Andean world more generally is that of the *kancha*, a rectangular enclosure with stone walls around it and a single entrance, grouping together one-room buildings intended for a single function. The *kallanka* was another typically Inca architectural structure, which consisted of an enormous rectangular building with a double-pitched roof supported by pilasters. The interior was not divided and so could accommodate large crowds when necessary, for example during major celebrations or ceremonies of a public nature.

It was in 1533 that the Spanish conquerors reached the city of Cuzco after a tiring march across the highlands. They plundered the buildings, stealing the gold and silver found there, and massacred a significant part of the population. In 1536, a bloody insurrection took place under the reign of *Manqo Inca*. The indigenous people besieged the city for several months and burned many neighborhoods. The city was rebuilt by the Spaniards using the best artists and craftsmen. They built sumptuous colonial palaces and very beautiful churches. An earthquake in 1650 destroyed many of these buildings because, unlike the Inca constructions, they had not been designed to withstand seismic tremors.

7.2 On the Hill of *Saqsaywaman*

During the Inca era, there was a rather remarkable architectural complex on the hill of *Saqsaywaman*. The construction of this site began around 1460 and was the work of the Inca *Pachakuteq*. On his death, his son *Tupaq Yupanki* continued the construction and it was undoubtedly *Wayna Qhapaq* who completed it at the beginning of the sixteenth century (Niles 1999). Opinions differ regarding the use of this construction, most often considered as a fortress. It may also have been a place where ceremonies were organized to glorify the Sun (Zuidema & Quispe 1973), a city of refuge, a temple dedicated to the god of lightning, or a barracks. In any case, it was a remarkable architectural complex which testifies to the Incas' skill in adjusting and fitting together enormous carved stones. It was especially the first of the three floors that was remarkable (Fig. 7.4).

The orientation of Cuzco in relation to this site is such that, during the June solstice, the Sun first illuminates *Saqsaywaman* before illuminating the city. *Saqsaywaman* had three towers aligned from east to west of which only the foundations remain. A first square tower, located to the west and below the central level, was intended to store provisions or house soldiers. The central

Fig. 7.4 Enclosure wall of the *Saqsaywaman* complex. Author's photograph

tower (*Muyuq Marka*) was circular and was perhaps used for water storage purposes. It had four floors and some walls were richly decorated with gold and silver, according to Garcilaso de la Vega, which implies that the tower may possibly have served as accommodation for emperors, or that it was a temple dedicated to the worship of the Sun. The third tower was to the east of the complex and could possibly have housed a weapons store. The fortress had three levels of bastions or enclosures surrounded by walls, with the lower level having the largest blocks of stone. The esplanade facing the bastions was called Chukipanpa ('Esplanade of the Royal Lances').

The fortress was the site of one of the last episodes of the conquest of Peru by the Spanish. In 1536, during the *Manqo Inca* uprising, it was when the Spanish attacked *Saqsaywaman* that Juan Pizarro (1511–1536), the youngest of the four brothers, was killed. This episode was recounted by an eyewitness to the battle, Pedro Pizarro (1515–1602) (the cousin of the four brothers), in his chronicle *Relación del descubrimiento y conquista del Perú* [1571] (1992).

The 'Festival of the Sun' or *Inti Raymi* was already celebrated by the cuzqueños before the arrival of the Spaniards. It was a traditional celebration that originally took place on the day of the winter solstice. The purpose of this celebration was to invoke the Sun for good harvests. In exchange, two llamas were sacrificed, one white and the other black. The viscera of these animals were examined by the priests to make predictions and auguries relating to the coming year. At the winter solstice, the Incas, surrounded by their court and adorned in their most beautiful finery, gathered in the center of Cuzco to celebrate the sunrise. These festivities took place near the *usnu* of the main square of Cuzco (*Hawkaypata*), at the Temple of the Sun, in the Inca palaces, and in the *Aklla Wasi* or temple of the 'Virgins of the Sun.' Currently, part of the ceremonies organized on the occasion of this festival take place on the esplanade of the *Saqsaywaman* fortress.

7.3 Strange Rocks

Four kilometers northeast of Cuzco, on the road to P'isaq, there is a sanctuary called Q'enqo Grande ('the labyrinth'), famous for its rather unconventional rocks (Figs. 7.5 and 7.6). These curiously cut rocks undoubtedly already served as a sacrificial altar at the time of the Chavín culture. The name comes from the snake-shaped receptacle, carved in stone by the Incas, where priests poured blood or *chicha* during religious celebrations.

Climbing to the top of the site reveals the flat surface where the ceremonies took place. Q'enqo Grande has an amphitheater-shaped structure composed of a large semi-circular wall decorated with niches cut into the rock. The wall marks out a vast space containing a rocky outcrop and a monolith sculpted with zoomorphic figures and linear motifs. In the upper part is the *usnu*, comprising two cylindrical monoliths placed vertically on a block of stone. A canal starts from a cavity near *usnu* and reaches an underground room known as the 'sacrifice room' (Fig. 7.6).

The Incas admired light shows at Q'enqo Grande called 'the awakening of the puma,' in which the puma appeared among the sacred animals with the condor and the snake (Van de Guchte 1990). Two gnomons approximately twenty-five centimeters high are located near and in the extension of a cave in the rock which is illuminated by solar light during sunrise at the time of the June solstice.

The cave has two entrances. It contains altars, niches for placing mummies, and ritual stairs. The cave is oriented in the southeast–northwest direction, according to the orientation of the solstices. At the time of the June solstice, around noon, sunlight enters the main cave and illuminates the altars and staircases, the latter materializing the transition between the underworld, the terrestrial world, and the celestial world. At the time of the equinoxes, the altar

Fig. 7.5 The unconventional rocks of the Q'enqo Grande sanctuary. Author's photograph

of the secondary cave is illuminated by the light of the sunrise. The opposite entrance to the cave, near the main altar, is oriented toward the sunset at the equinoxes. More details on this site can be found in Gullberg (2009, 2020).

7.4 The Sacred Valley of the Incas

The sacred valley of the Incas is formed by the valley of the Urubamba river (its upper course is called Vilcanota and its lower course Ucayali) between P'isaq and Ollantaytambo. After this town, the valley narrows and becomes a deep gorge covered with tropical vegetation. This region was used for agriculture from very early on. According to Spanish chroniclers, the valley was described as sacred by the Incas because there was very fertile soil, thermal springs, salt pans to collect salt, and also abundant pure water. Medicinal plants also grew there, used to treat altitude sickness (Molinié-Fioravanti 1982).

On a road leading to P'isaq are the ruins of Puka Pukara (the 'Red Fort') at an altitude of 3650 m (Fig. 7.7). This is in a strategic position for maintaining access to Cuzco. Puka Pukara may have been a hunting lodge or a stopping

Fig. 7.6 The 'sacrificial room' of the Q'enqo Grande sanctuary. Author's photograph

place for travelers, but it was certainly also a guard post. It features a series of rectangular constructions with terraces, interior squares, aqueducts, and fountains. Puca Pucara appears to have been a residence of *Pachakuteq* (Bauer 1998). The doors and corridor of this site are orientated according to the cardinal points and also according to the equinoxes (Gullberg 2009, 2020).

About 8 km from Cuzco, we find the site of Tampumach'ay (or Tambomachay) ('place of rest'). The Tampumach'ay cave is associated with a platform and a staircase, the platform facing the sunrise during the December solstice. The cave is also oriented relative to the solstices. The site has a dominant position over Tampumach'ay, Puca Pucara, and part of the Cuzco valley.

The sacred springs are sometimes called El Baño del Inca ('the Inca's bath'). This may have been a second home of the emperor. It has three levels of *andenes* marked out by carefully constructed walls. Of particular interest is the Baño de la Nusta ('the princess's bath') formed by two aqueducts cut into the rock which feed a ceremonial fountain. The site runs on a complicated hydraulic system with underground pipes. We can also see the 'sacred' water used by the Inca priests as it emerges from the rock. This water had the reputation of being the milk of the *Pachamama*. According to local beliefs, it would bring fertility,

Fig. 7.7 The Puka Pukara site, a strategic location guarding access to Cuzco. Author's photograph

eternal youth, and beauty to those who drank it. The noble Incas waited for sunrise on the upper terrace to perform the ceremonial 'water' rite (Figs. 7.8 and 7.9).

The Watanay valley was densely populated and had villages, cultivated terraces, granary complexes for storing provisions, temples, and properties belonging to members of the aristocracy in its foothills. The sovereigns had quite luxurious country houses in the Vilcanota valley, north of Cuzco, between P'isaq and Machu Picchu. One of the most elegant buildings was built by *Wayna Qhapaq* near the current center of Urubamba.

Pachakuteq celebrated his victory over the Cuyos by building a palace at P'isaq. This site is located 17 km northeast of Cuzco and dominates the Vilcanota River from the top of its terraces. This complex is characterized by solstitial and equinoctial alignments (Gullberg 2009, 2020). The site has three parts: the fortress, the sanctuary or Temple of the Sun, and the city. The fortress offers a superb view of the surrounding valley. Certain stones, which form the base of this building, are gigantic and very nicely assembled. The sanctu-

Fig. 7.8 The site of Tampumach'ay may have been one of the emperor's secondary residences. Author's photograph

ary housed religious ceremonies, but Inca dignitaries and religious authorities would certainly have stayed at the P'isaq site.

The main building has a straight wall on one side and a curved wall on the other. It is associated with a rocky outcrop which occupies the center. Like other major Inca centers, it is home to an *intiwatana* ('anchoring place of the Sun'). In another construction to the east of the *intiwatana*, there is an orientation platform (*usnu*) intended for observing the stars, and in particular the Sun. It is oriented toward the southeast horizon at an angle of 114°, corresponding to sunrise at the time of the December solstice. On the west side of the carved rock, we find a representation of the upper part of a Chacana cross. The lower part of this cross is visible when proper alignment is achieved and is formed by the shadow cast by the sculpture during the equinox (Gullberg 2009, 2020) (Figs. 7.10, 7.11, 7.12, and 7.13).

The sector containing the *intiwatana* was reserved for the political and religious elite, and the adjacent complex (Pisaca) onto which it opened was characterized by narrow corridors and three portals which made it possible to control access and maintain a certain discretion within these careful construc-

Fig. 7.9 In Tampumach'ay, we can see the 'sacred' water used by the Inca priests as it emerges from the rocks. Author's photograph

tions. The Qalla Casa sector, on the other hand, was closely associated with terraces and agricultural production.

The Ollantaytambo site is located 90 km from Cuzco in the Urubamba river valley. If we are to believe the chronicler Garcilaso de la Vega [1609] (1969, 2000), the site of Ollantaytambo was built by the Inca *Wiraqocha*. The name means 'resting place of Ollanta,' named after the warlord Ollanta, who helped the *Sapa Inca Pachakuteq* conquer the kingdoms of the northern coast of Peru. To thank him for his exploits, the prince said he would grant him any wish. Ollanta asked for the hand of his daughter Kusi Quoyllur with whom he was in love. But the prince refused because he was not of noble ancestry. The princess, refusing to marry another man, chose to remain a virgin, devoting herself to the worship of the Sun. Actually, the story found its epilogue after the death of *Pachakuteq*.

Pachakuteq actually developed the Ollantaytambo region after his victory over the Tambos. The place was later improved by *Wayna Qhapaq* and *Manko Inca* (Niles 1999). An important feature of this site is the magnificent staircase which climbs the mountain. Here, the terraces of Pumatillis have an orientation

Fig. 7.10 The P'isaq site is located on the right bank of the Rio Vilcanota. Author's photograph

of approximately 114° and thus face the sunrise during the June solstice. Many pilgrims would have used this staircase to observe the sunrise during the festival of *Qhapaq Raymi* (December solstice). The Ollantaytambo Sun Temple was essentially destroyed by the Spanish and only a few monoliths of the foundation walls remain. From this, it is difficult to determine whether a preferred direction was favored during its construction. Opposite Ollantaytambo, the Pinkuylluna mountain is aligned with the sunrise at the June solstice. The characteristic shape of this mountain was believed to represent the figure of Tonupa, the messenger of the god Wiraqocha transformed into stone (Salazar & Salazar 2014).

Dominated by the fortress, the village with its narrow streets constitutes a fine example of Inca architecture. It was divided into *kanchas*, blocks of houses with a single entrance opening onto a courtyard. The village has two parts separated by the Patacancha stream, a tributary of the Urubamba. The eastern part (Qosko *Ayllu*) is formed by the current village, while the western part (Aracama *Ayllu*) included temples, an astronomical observatory, ceremonial areas, food stores, terraced fields, and a fortress. On the top of the mountain opposite the

Fig. 7.11 P'isaq. General view of part of the site. Several terraces can be seen in the foreground. Author's photograph

main complex of Ollantaytambo, one can see the ruins (Pinku Lluna) of what was undoubtedly a warehouse for agricultural products (Figs. 7.14, 7.15, and 7.16).

Ollantaytambo was a major religious center, but it was also a formidable fortress (Protzen 1993). It was in Ollantaytambo that *Manqo Qhapaq* came to retire after the defeat he had suffered at *Saqsaywaman*. In 1536, Hernando Pizarro attempted to capture the rebel leader but failed, and the conquistadors had to withdraw. However, the Spanish later returned with reinforcements, and *Manqo Qhapaq* had to flee, before finally being killed by the Spanish.

7.5 Places of Pilgrimage

In addition to the Sun and the Moon, the people of the Andes revered a wide range of places and objects, including mountains, lakes, and rocks, because they believed these had a sacred character. Some of these *wakas* were small, but others required considerable maintenance and were frequented by pilgrims,

Fig. 7.12 P'isaq. Detail of the very elegant buildings located at the site. Author's photograph

sometimes from distant lands. In some places, offerings of animals or agricultural products were made. The peoples conquered by the Incas were not required to abandon their beliefs but, in addition to the local gods, they had to honor the Sun and the Moon, the official deities of the empire (Malville 2010).

The Temple of the Sun or the Golden Enclosure

The most famous sanctuary of the empire was the Temple of the Sun (*Templo del sol*) or *Qorikancha* ('House of Gold') located in the center of the city of Cuzco, already discussed above. It was initially called *Inti Kancha* ('Enclosure of the Sun'), but the date of its construction has not been determined. According to some, *Manqo Qhapaq* commissioned it, but according to others, it was the *Sapa Inca Pachakuteq*. This sacred enclosure, overflowing with gold, was built with beautifully cut stone, an art in which the Incas excelled. It was made up of a series of buildings dedicated to different deities. Together with the

Fig. 7.13 P'isaq. The main building has a straight wall on one side and a curved wall on the other. It is associated with a rocky outcrop which occupies the center. Like other major Inca centers, it houses the *intiwatana* and an altar dedicated to worship. Author's photograph

neighboring square and several terraces, it constituted a unique architectural ensemble, visible from afar (Lehmann-Nitsche 1928).

This temple, located near the confluence of the two main rivers supplying Cuzco, had high straw-thatched roofs with four slopes, like most of the palaces in Cuzco. What set it apart from the others was the large amount of gold it contained. Thus, one could admire in particular a gold fountain in the middle of the main square and several exterior walls covered with plaques of the same metal. A detailed description of this palace was given in particular by Pedro Cieza de León [1553] (1984) (Figs. 7.17, 7.18, 7.19, and 7.20).

Opposite the temple was Intipanpa, the 'Garden of the Sun,' which was decorated with solid gold figurines, plants, fruit, and animals. The different wings of the *Qorikancha* were dedicated to the main deities, namely the Sun, the Moon, the stars, the thunder, the rainbow, and the creator god *Wiraqocha*. The central idol was a representation of the Sun god in gold. It had the form of a young man (*the P'unchaw*). Similarly, the representation of the Moon goddess

7 Proximity of the Gods and Sacred Places

Fig. 7.14 Ollantaytambo. Partial view of the site. Author's photograph

had the form of a woman. The *Qorikancha* was undoubtedly the most sacred place in the entire empire. In particular, it contained the mummies (*mallki*) of the Incas *Pachakuteq* and *Wayna Qhapaq*. As this temple contained large amounts of gold, it was plundered by the Spanish invaders, who took away most of its riches.

The *Qorikancha* was destroyed before 1560. After the fall of the Inca empire, the monastery of Santo Domingo was built on the site of the temple, and only part of the foundations and walls built of cut stones remain. Excavations carried out near the site revealed pottery dating from the period 1000–1400 CE, showing that the place was occupied before the advent of the Inca empire.

Pachakamaq, The Creator of the World

Pachakamaq was located on the coast a short distance (about 30 km) south of the current city of Lima. The majestic ruins of this site bear witness to the important role this ceremonial site played before the arrival of the Spanish. Built in adobe, this was one of the main places of worship in pre-Hispanic Peru.

Fig. 7.15 Ollantaytambo. Detail showing the construction of a wall. Author's photograph

The site was excavated by archaeologists like M. Uhle and Julio Cesar Tello Rojas and more than fifty local architectural structures are currently known.

It was certainly a popular place of pilgrimage. People came to worship the effigy of *Pachakamaq* ('the creator of the world'), made offerings, and listened to the oracles. After the incorporation of this region into the Inca empire, the temple remained a very famous *waka*. The Incas built a Temple of the Sun on a hill facing the sea and it seems that even emperors came to hear the oracular predictions. Near the temple, there was also accommodation for priests, for pilgrims, and also for 'chosen women,' whose vocation was to serve not only the gods but also the emperor.

The origins of this site are poorly known. Excavations have shown that the place was already frequented as a pilgrimage center during the era of Wari supremacy. A symbolic representation of the god *Pachakamaq* was found in 1938 in the Templo pintado ('Painted Temple'). It is a slim wooden sculpture featuring anthropomorphic representations accompanied by felines, birds, and plants. The lower part of this idol includes a symbolic representation of the three parts of the Andean universe, namely, *Urin Pacha, Kay Pacha*, and *Hanan Pasha*. During excavations of the site, numerous offerings were found, mainly

Fig. 7.16 Ollantaytambo. On the top of the mountain opposite the village, there is an enormous complex which was undoubtedly a warehouse for agricultural products. Author's photograph

stones carved in the shape of ears of corn, peppers, and potatoes that pilgrims buried while invoking *Pachamama* (Pozzi-Escot et al. 2017).

The Islands of the Sun and the Moon

According to the myth told by Spanish chroniclers (see Sect. 5.1), the Sun was born on an island in Lake Titicaca. The first couple from which the whole human race was born also originated on this island. At the time of the arrival of the Spanish, the island had a series of temples which formed a religious complex on the south side of the lake and an important temple built near a sacred rock (Dearborn et al. 1998). According to the account given by Francisco Pizarro, who probably visited this place at the end of 1533 or the beginning of 1534, and subsequent chroniclers including Bernabé Cobo, there were in fact two sacred islands: one was called Titicaca ('Island of the Sun') and the other Coati ('Island of the Moon'). There is also a sacred rock, called Titikala, at the very

Fig. 7.17 Only part of the foundations and walls of the *Qorikancha* remain. Author's photograph

place where, according to legend, the Sun and the Moon were born. There were various temples and a structure for 'chosen women' on the two islands.

In accordance with Inca cosmology, the Sun and the Moon constituted a sacred couple. The Sun was identified with the king and masculinity while the Moon was associated with the royal wife and femininity. The royal consort had a mother–daughter relationship with the Moon, similar to the unwavering relationship the emperor had with his father the Sun.

The worship of the Sun by the Incas is unanimously recognized in all the chronicles of the day. During religious ceremonies, the emperor was the sole mediator between the people and higher cosmological forces. This relationship between the Sun and the sovereign was reinforced by the fact that the *P'unchaw* of the *Qorikancha* contained substances from the dried hearts of deceased sovereigns.

Fig. 7.18 After the fall of the Inca empire, the monastery of Santo Domingo was built on the site of the temple. Author's photograph

7.6 Machu Picchu: At the Heart of the Sacred and Observation of the Sky

The Mysterious City of Vilcabamba

Machu Picchu (or Machupijchu, in Quechua) is one of the best-known sites in South America (Figs. 7.21 and 7.22). It is an ancient Inca city from the fifteenth century, perched on a rocky promontory which connects Machu Picchu ('Old Mountain'), on the one hand, and Huayna Picchu ('Young Mountain'), on the other hand. Located in an exceptional and spectacular setting, it consists in the admirably preserved ruins of a prestigious pre-Columbian civilization. This city would have served as the residence of the emperor *Pachakuteq*, but the access road and various constructions on the site indicate that this place would also have been a religious sanctuary. Machu Picchu was apparently the site of festivities and religious ceremonies during the dry season in which the aristocratic families of Cuzco doubtless took part.

Fig. 7.19 The *Qorikancha*. Partial view of what remains of the interior walls. Author's photograph

The site is located in the east of the Andes mountains, approximately 130 km from Cuzco (Urubamba province). It sits at an altitude of 2470 m. Below the ruins, at the foot of a 600-meter-high cliff, the Vilcanota-Urubamba river describes a large loop. The place is accessible on foot via the 'Inca Trail' or by bus from Aguas Calientes, a village that can only be reached by train from Ollantaytambo.

Several Spanish chroniclers mentioned the existence, in the jungle north of Cuzco, of a mysterious city which bore the name of Vilcabamba and which was apparently not explored by the conquistadors. Travelers in the nineteenth century, such as the Frenchman Charles Wiener (1851–1913) and the Italian Antonio Raimondi (1826–1890), actually confirmed the existence of a site dominating the Rio Urubamba, but without having explored it themselves. Revealed to the world in 1911 by Hiram Bingham (1948, 2003), a professor at Yale University whose expedition to Peru was partly financed by the *National Geographic Society*, this astonishing site has preserved part of its mystery: When exactly was it built? What was it used for? When and why was it abandoned?

Given the absence of written records relating to Machu Picchu, our knowledge of this site comes from similar architectural structures observed elsewhere

7 Proximity of the Gods and Sacred Places

Fig. 7.20 The *Qorikancha*. Detail of the interior structures. Author's photograph

in the empire, what we know about the religious beliefs and practices of the Incas, and logical deductions made from observations made *in situ*. According to archaeologists, the city was apparently built by the emperor *Pachakuteq* (1438–1471) around 1440 and it may be that he erected it in homage to the creator god *Teqsi Wiraqocha*. The population of this extraordinary little town was between a few hundred and fifteen hundred inhabitants, probably belonging to the religious and political elite. On the death of the emperor *Pachakuteq*, the site became the property of his *panaka*, a situation which continued under the reigns of subsequent emperors, namely *Tupaq Yupanki* and *Wayna Qhapaq*.

The civil war over the succession of the throne which broke out in the Inca empire in 1531–1532 and also the arrival of the Spaniards shortly afterwards undoubtedly spelt the end of the occupation of Machu Picchu. The nobles would have returned to the court in exile in Vilcabamba following the appeal of *Manqo Inca* in 1536. This situation favored the departure of peasants who had been forcibly deported to these regions and they would also have returned to their lands of origin. The Machu Picchu region would then have suffered the Spanish *Encomienda*. It cannot be excluded either that the site would have become a victim of the 'extirpators of idolatry,' a kind of forced evangelization.

Fig. 7.21 General view of Machu Picchu with the mountains in the background. Author's photograph

In any case, it was gradually abandoned and fell into relative oblivion for several centuries, even though several travelers may have visited it.

Following the 'Inca Trail'

The site of Machu Picchu was a place of residence of the emperor, but it was also intended for religious practices. It is unlikely, given the way it was laid out, that it served defensive purposes. In the southern area of the site, there are numerous terraces which had an agricultural vocation, the work being carried out by peasants from different regions of the empire. This local workforce would have supplemented the agricultural production of the densely populated neighboring valleys. The retaining walls were designed to use irrigation without suffering the damage that could be caused by water. In this way, these agricultural lands could be used to cultivate corn (perhaps for sacrificial purposes), along with potatoes and various kinds of vegetables. The development of the terraces would have been linked to the hydrological cycle, and that in turn would have been related to the glorification of mountain deities (Zuidema 1978).

7 Proximity of the Gods and Sacred Places

Fig. 7.22 Machu Picchu is located in the east of the Andes, approximately 130 km from Cuzco (Urubamba province). It sits at an altitude of 2470 m. Author's photograph

The urban area, to use the archaeologists' term, was where most religious or civil activities took place. In addition to the public areas, there was also a sacred area and an area for the residences of priests and nobles. In accordance with the dichotomous system governing the organization of Andean society, archaeologists distinguish between an upper district and a lower district.

The site of Machu Picchu was undoubtedly accessible to selected pilgrims who reached it by taking the path now called the 'Inca Trail.' This trail through the Cordillera Vilcabamba has existed for more than five hundred years and allows the visitor to discover various remains (houses, stairways, tunnels, etc.) as well as the ruined sites of Runru Raqay, Sayaq Marka, Phuyupata Marka, and Wiñaywayna, accessible only on foot.

Many of the constructions in Machu Picchu are made from perfectly squared stones according to the custom of the Incas, but some walls were built using unfitted stones. However, all these buildings have excellent resistance to earthquakes.

Fig. 7.23 On the Sacred Square, rather small in size, you can admire the Temple of the Three Windows, a building with a marked ceremonial character from which there is a stunning view of the surrounding mountains. Author's photograph

Machu Picchu and Sacred Geography

If we are to believe Reinhardt (2007), the perception of Machu Picchu cannot be dissociated from the so-called sacred geography of the Incas, and many buildings on this site have ritual and religious significance, such as the Temple of the Three Windows (Fig. 7.23), the Astronomical Observatory (Fig. 7.24), the Temple of the Moon, the *Intiwatana* (Fig. 7.25), the Sacred Rock (Fig. 7.26) and so on (Dearborn & Schreiber 1986). However, we have little information about the occupation of the site during the Inca era (Salazar 2004). The sacred area was dedicated to the god *Inti*, the Sun god, and to *Wiraqocha*, the creator god. There we find the *Intiwatana* and the Temple of the Sun, also called the *Torreón*, a sort of conical tower, built on a large rock (Fig. 7.27). This rock shelters a cavity (Fig. 7.28) which was perhaps used to deposit the mummies of the missing emperors (Waisbard 1976).

In the Andes, and in the Cuzco region in particular, the mountains were particularly important in the religious life of the Incas. They embodied powerful traditional deities who had to be regularly propitiated. They were invoked to increase the fertility of herds and, in particular, camelids like llamas and

Fig. 7.24 The astronomical observatory is located on a rocky promontory accessed from the Main Temple by a staircase with seventy-eight steps. Author's photograph

alpacas, to influence the weather, to bring protection from wild animals, enemies, or disease, to promote trade, and so on. Very high mountains, especially those covered with snow, were particularly glorified, but lower peaks were also honored. This is not surprising given that mountain peaks are the site of many weather phenomena, sometimes violent (rain, snow, clouds, lightning, thunder, etc.). They are also home to many wild animals which, according to popular beliefs, belong to them (pumas, bears, birds of prey, etc.). The same goes for llamas and alpacas, which would have played a crucial role in the local economy, given that mountain deities were reputed to be responsible for their fertility. In this context, honoring the mountain gods took on a major importance, and the priests who embodied the worship given to these superior beings, benefited from particular prestige because they contributed to the prosperity of the people, preventing disease from spreading, and they were the basis for the development of prosperous trade.

Mount Salqantay, which rises to 6271 m above sea level, was a particularly revered mountain that dominates the Cuzco and Machu Picchu region.

Fig. 7.25 The *Intiwatana* is a stone table whose corners are oriented towards the cardinal points and which features a kind of spur, all carved from a single perfectly polished stone. Author's photograph

Fig. 7.26 Another rock on Machu Picchu, called the Sacred Rock, was carved to imitate the shape of a mountain. The mountain in question is perhaps Mount Yanantin, which was honored in the Machu Picchu region, but it could also be, as has been suggested, Pumasillo mountain. Author's photograph

Fig. 7.27 Machu Picchu. The Temple of the Sun or *Torreón* is built on an imposing rock. Below it there is an opening in which mummies may have been placed. Author's photograph

Another nearby peak is Mount Ausangate (6372 m), which was also respected and made the subject of various celebrations, rites, and offerings.

The Spanish chronicler Cristóbal de Albornoz [ca. 1582] (1984) was undoubtedly one of the first to reveal the cult paid to these mountain peaks. However, these were considered by the people as irascible deities who had to be appeased to avoid unpleasant surprises, because their power was immense. This power was not only manifested locally, even though there was a line of demarcation separating the zones of influence of the deities embodied by Ausangate and Salqantay. Thus some considered that the Ausangate mountain could extend its power to the city of Puno on Lake Titicaca!

From the Main Temple to the *Intiwatana*

The Main Temple is a structure open to the south, with a huge stone slab as altar on the north side. This building opens onto the Sacred Square, a rather small area from which one can admire the Temple of the Three Windows, a building clearly destined for ceremonial purposes and giving beautiful views

Fig. 7.28 Machu Picchu. Space cut from the rock under the Temple of the Sun. This cavity was perhaps a mausoleum for mummies. Author's photograph

of the surrounding mountains (especially towards the Pumasillo massif) and below, down to the Urubamba River. This place, connected to the *Intiwatana* by a staircase, was very probably dedicated to the cult of water (Fig. 7.23). An *Intiwatana* can be found at several Inca sites, but the best preserved is undoubtedly the one at Machu Picchu. It is also one of the most important structures on the site, accessed by a stairway with 78 steps from the Main Temple.

The *Intiwatana* is a stone table whose corners are oriented towards the cardinal points and which features a kind of spur, all carved from a single perfectly polished stone. This stone, 1.80 m high and cut from a single block, played an important role in establishing the calendar. It bears this name by analogy with the name attributed in 1856 to a similar stone at the site of Ollantaytambo, where there is also a Temple of the Sun. The P'isaq site includes a stone of this type, too (Fig. 7.25).

It was in all likelihood an astronomical observatory from which the priests observed the movement of the stars, and the Sun in particular, in order to determine the key dates of the agricultural calendar. If we observe the shadow cast by the rocky spur when the Sun shines, we see that it is the shortest on the north side at the time of the summer solstice on 21 December, and the longest on the south side at the winter solstice on 21 June. It was on the latter date, according to the Incas, that the Sun was furthest from the Earth since the shadow was the longest. Therefore, to prevent the day star from continuing to move away from the Earth and leaving the people to die of cold and hunger, the Incas attached it to a stone pillar. In this context, the winter solstice was a time of fear while the summer solstice was instead marked by rejoicing.

The positioning of this rock is interesting with regard to the four cardinal directions. On the east side, we find the Veronica massif, the Sun rising behind its highest peak at the time of the equinoxes. To the west, we see the Pumasillo massif, the Sun setting behind its highest peak at the time of the December solstice. The Salqantay massif, not observable from the *Intiwatana* but visible from Huayna Picchu, is to the south, its highest summit located at an azimuth of 180°. Huayna Picchu is the mountain located to the north. The *Intiwatana* therefore occupies a central position where astronomical observations are concerned and when we consider the mountains particularly honored within the framework of a sacred geography.

Regarding solar observations, at the time of the equinoxes, the Sun sets behind Cerro San Miguel. At the June solstice, it rises behind Cerro San Gabriel, part of the Veronica massif. At the December solstice, the day star seems to emerge from the Urubamba River. It is also possible to observe the Southern Cross moving near the peak of Machu Picchu. The *Intiwatana* is therefore located in a privileged location for carrying out solar observations.

It should be noted that the shape of the *Intiwatana* stone reproduces the mountain located in the background, namely Huayna Picchu. A rock, also carved to replicate the Salqantay mountain in the background, is visible at the top of Huayna Picchu. It may be that the *Intiwatana* symbolized the 'spirit' of the mountain on which it was built. It is not strange in fact to consider such a hypothesis because the Incas also honored a carved stone representing the shape of the Huanacauri mountain. Another rock on Machu Picchu, called the Sacred Rock, was cut in a similar manner, the mountain concerned probably being Mount Yanantin, which was honored in the Machu Picchu region, or perhaps the Pumasillo mountain.

As we know, the Southern Cross played an important role in the astronomy of the Incas. Indeed, there is a relationship between this constellation and the Salcantay mountain: observed from Machu Picchu, the Southern Cross rises to the east and sets to the west of Mount Salcantay, and at its highest position in the sky (to the south), it is above this mountain.

Not far from the Southern Cross, we find the stars α and β Centauri, identified with the eyes of the Llama. According to the Inca belief system, the Llama constellation was closely involved in terrestrial life, and agricultural life in particular, because it was present in the sky before and during the rainy season. In the Cuzco region, it was thought that the llamas were under the protection of the Ausangate mountain. Generally speaking, the animals appeared as dark clouds in the sky, and were supposed to come from Earth and enter the sky via the top of a mountain.

Near the Llama on the celestial vault and on the same side of Salcantay when the Southern Cross culminates above the summit of this mountain, we find the constellation of the Fox (*Atoq*), an animal considered to be the assistant of the mountain deities. The Sun rises in the Fox constellation at the time of the December solstice, that is to say, at the start of the rainy season. At this time, seen from Machu Picchu, the Sun appears in the morning above the Urubamba River and seems to rise towards the celestial river (the Milky Way).

A place in Machu Picchu admirably carved into the rock and accessible by stairs is the Temple of the Sun or *Torreón* (Fig. 7.27). Offerings were placed in the hollows of the holes dug into the walls of this building. The opening made in the adjoining rock has sometimes been considered a royal tomb, even though no bones have ever been found there (Fig. 7.28). The upper part of this structure was used as a sacrificial altar, but the place may also be the site of astronomical observations, and in particular, solar observations at the time of the June solstice. During the winter solstice, the light of the rising Sun falls on a straight line engraved in the middle of the tower. An opening also allowed the observation of the Pleiades at this same time of the year, important observations in relation to weather forecasting and the abundance of crops (Dearborn & White 1983).

On the other side of the mountain at the same time, we find the constellations of the Toad (*Hanp'atu*), the Snake (*Mach'aqway*), and the Tinamou (*Yutu*). According to Andean beliefs, these three constellations were associated with water and fertility. The Toad constellation appeared in the sky at the time of the rainy season (Fig. 7.29), while Tinamou represented the mountain gods and was also associated with the wet season and the rainbows present in the sky at that time of year.

Fig. 7.29 Anuran amphibians during reproduction. The Toad constellation appeared in the sky during the rainy season. Photograph taken by the author at the Larco museum in Lima

Under the Temple of the Condor, a temple which owes its name to a condor engraved on a flat stone placed on the ground, there is a small cave called the Intimachay. This is a viewing location for the December solstice. The front wall of the cave has an opening through which the light of the rising Sun enters during the few days preceding and following this special moment of the astronomical year. It then directly hits the rear wall of the cave (Ziólkowski et al. 2013).

7.7 Lake Titicaca and Surrounding Areas

Lake Titicaca is the largest lake in Latin America and the largest lake higher than 2000 m above sea level (Fig. 7.30). It is in fact made up of two lakes: the small one, or Huiñaimarca, with its eleven islands, and the large one, Chucuito, with twenty-five islands. Thanks to its wealth of water and fish, the lake has allowed many civilizations to make a living in the region. Lake Titicaca is considered

Fig. 7.30 View of Lake Titicaca. Author's photograph

sacred because, according to legend, it was from its waters that *Manqo Qhapaq* and his sister-wife *Mama Oqllo* arose to found the kingdom of *Tawantinsuyu* (see Sect. 5.1).

In Peru, the *altiplano* is also called the *puna*. These are vast plains with the peaks of the Cordillera Real in the background. It is a paradise for vicuñas, alpacas, and llamas, but the *campesinos* (or local farmers) generally live in relative poverty at altitudes between 2800 and 4500 m above sea level. Their adobe or stone houses stand near fields of crops such as corn, quinoa, beans, and different varieties of potatoes (Fig. 7.31).

The largest city in the region is the town of Puno which was founded in 1668 near a silver mine called Laykakota. The floating islands of Puno Bay are inhabited by the Uros peoples. According to some accounts, the Uros took refuge on rush islands to escape the Incas' demands for tributes. These people

7 Proximity of the Gods and Sacred Places   127

Fig. 7.31 Typical constructions of the Peruvian *altiplano*. Author's photograph

Fig. 7.32 The Uros live on layers of rushes several meters thick. They still use traditional boats. Author's photograph

Fig. 7.33 The manufacture of fabrics is one of the traditional activities of the Uros peoples. Author's photograph

still live on layers of rushes several meters thick which must be renewed regularly. Their boats are also made of rushes and used for fishing and transporting passengers. A few years ago, these people, who refused any integration into Peruvian society, still lived solely by hunting and fishing, and fed in particular on young totora shoots (Figs. 7.32 and 7.33).

8

Measurement of Time, *Seq'es*, and Associated Rites

8.1 Synodic Period and Tropical Year

Astronomers define several different periods of revolution of the Moon. These include the sidereal period, which corresponds to two passages of the Moon[1] at the same position in the sky in relation to the stars, the synodic period or lunation, which corresponds to the return of the Moon to the same position in relation to the Sun, and the tropical period, which corresponds to two successive passages of the Earth's natural satellite at the same position in relation to the vernal point.

The true solar day corresponds to the time interval between two consecutive passages of the center of the Sun across the meridian at the given location. This is not actually constant over time, but fluctuates annually due to the obliquity of the ecliptic and also the elliptical nature of the Earth's orbit. This is why, rather than the true Sun, we prefer to consider a fictitious 'mean' Sun which travels towards the east in a uniform motion along the equator rather than along the ecliptic, and which can be used to define the mean solar day, that is, the time interval between two successive passages of the mean Sun at the meridian of the given location.

The tropical year is the time interval between two consecutive passages of the Sun at the vernal point. It differs from the sidereal year, which is the time

---

[1] For more details on the astronomical notions discussed in this chapter, the reader is referred to Appendix B.

interval between two consecutive passages of the Sun through the same point of its apparent orbit on the ecliptic, therefore in relation to the stars, the reference point for longitude being the vernal point.

8.2 The Motion of the Sun and the Rhythm of the Seasons

We know that, during the year, the Earth follows an elliptical trajectory around the Sun in accordance with Kepler's laws, the Sun occupying one of the foci of this ellipse (see, for example, Biémont, 2000). In the astronomy of appearances, we may consider that it is not the Earth which moves around the Sun, but that it is the Sun which moves around the Earth, a point of view which prevailed before the work of Copernicus and which allows an easy description of the annual progression of the seasons. Indeed, the annual translational motion of the Earth around the Sun (or the apparent motion of the Sun in the sky) allows us to define the plane of the ecliptic, whose inclination relative to the plane of the Earth's equator is responsible for the phenomenon of the seasons.

During an apparent revolution of the Sun around the Earth, the declination of the former varies between two extreme values corresponding to the summer solstice around 21 June and the winter solstice around 21 December in the northern hemisphere. The declination of the Sun goes to zero at the points Υ and Υ', which define the spring equinox around 21 March and the autumn equinox around 21 September, respectively.[2] These four privileged moments of the astronomical year correspond to the beginning of the seasons. It is important to note that the winter solstice in the southern hemisphere occurs around 21 June and the summer solstice around 21 December since the seasons are reversed compared to the northern hemisphere. Similarly, the spring equinox occurs around 21 September and the autumnal equinox around 21 March.

A classic way, known since ancient times, of observing the passing of the seasons is to use a sundial and look at the shadow cast by a vertical rod planted in the ground, which constitutes the gnomon. Outside of the tropics, the length of this shadow will vary throughout the seasons. It will take a minimum value when the Sun is highest in the sky (at the time of the summer solstice) and a maximum value when the Sun is very low on the horizon (the winter solstice).

At the time of the spring and autumn equinoxes, the day and night have equal lengths at every point on the globe. The tropical year corresponds to the time interval separating two successive spring equinoxes. Regarding the

---

[2] The letter Υ is the symbol for the constellation of Aries into which, conventionally, the Sun enters during the spring equinox.

illumination of the Earth by the Sun, the dividing line between day and night passes through the poles. At the summer solstice, the Sun is at the zenith at noon for points located on the Tropic of Cancer. The regions located near the North Pole are permanently illuminated and we speak of the Midnight Sun. At the winter solstice, the Sun is at the zenith at noon for regions located on the Tropic of Capricorn. At this time, the area close to the pole is no longer illuminated.

The Sun rises due east and sets due west at the equinoxes. At noon, it cuts the arc of a circle passing through the zenith and due south of the given location, that is to say the meridian. During the year, the azimuth of the Sun at sunrise (and at sunset) varies between two extreme positions for a point of fixed latitude.

Now suppose we observe the rising or setting of the Sun on the horizon from a fixed point, which may be an astronomical observatory or a tower located on the top of a mountain. If, in the northern hemisphere, the observer looks at the Sun at the time of the spring equinox, he will notice that the Sun rises opposite him. He will then see the sunrise moving gradually to the left and, at the time of the summer solstice, he will observe the sunrise furthest to the left. In the following weeks, the Sun will gradually move to the right to rise again in front of the observer at the time of the autumn equinox. This motion of the sunrise will continue to the right to reach the extreme point around the winter solstice. The solstices and equinoxes will therefore appear at particular points on the horizon that are quite easily identifiable.

The times at which the Sun reaches its zenith depend on the latitude. These two passages grow closer as the latitude increases. This passage at the zenith can easily be observed from an underground observatory which allows sunlight to enter through an opening in the ceiling. The nadir is the point in the sky vertical to the observer but looking down. It is therefore located opposite the zenith.

8.3 The Inca Calendar

8.3.1 The Gregorian Reform and the Inca Calendar

In Western countries, we mainly use the Gregorian calendar for counting time. The Julian calendar which was reformed by Julius Caesar in 46 BCE was in use in Western countries until 1582 when the reform imposed by Pope Gregory XIII took place. On this date, a correction of around ten days was applied and we passed without transition from 4 to 15 October. The leap year system was also reformed and the calendar became Gregorian. In this Gregorian calendar, years whose number is divisible by one hundred are no longer considered leap years, with the exception of those which are multiples of 400. Thus 2000 was a leap year in the Gregorian calendar, while 1900 was not. Being a Catholic country, Spain quickly implemented the reform of Pope Gregory XIII. In Peru, the calendar became Gregorian in 1584, and in that year it jumped directly from 4 to 15 October. Dates prior to 1584 and relating to the history of Peru must therefore be corrected by around ten days. Thus, if a document dated 1580 mentions mid-March as a reference, it should be considered that this date actually corresponds to the end of the month in the Gregorian calendar.

The information we have concerning the calendar and the measurement of time among the Incas comes mainly from Spanish chroniclers and is unfortunately often fragmentary. There are even disagreements between the chroniclers. Among these we should mention the names of Pedro Sarmiento de Gamboa [1572] (1942, 2007), Juan Diez de Betanzos [1551] (1987), Cristóbal de Molina [ca. 1575] (1989), Polo de Ondegardo [1585] (1916), Bernabé Cobo [1653] (1956), Felipe Guámán Poma de Ayala [1615] (1936, 1980), and Garcilaso de la Vega [1609] (1945, 1969, 2000) (see Sect. 3.3).

As the Incas did not use writing, they did not leave us any simply accessible source of information about their calendar. The only source we have may possibly be the *khipus*, assemblages of cords whose colors, numbers of knots, and shapes convey information of different kinds. However, at the present time, their approach to the measurement of the time remains largely inaccessible to us (see Sect. 3.4), even though the work of contemporary researchers such as A.F. Aveni, G. Urton, or R.T. Zuidema has led to some progress in this area.

The Inca calendar was inspired by pragmatic considerations of an agricultural type (Itier, 2010), much more than by a pronounced taste for astronomy or the measurement of time. It was indeed vital for these people to know when to sow or harvest corn or potatoes, when to shear the llamas, when the rainy season was going to begin, etc. These dates, important for the life of the community, were determined mainly by solar observations, but also by stellar observations, as we will see later. It has been established, on the basis of the texts of different chroniclers, that the Incas carefully observed the movements of the Sun in the sky and marked certain sunrises and sunsets, notably at the equinoxes and solstices, by the presence of pillars, usually built on the hills surrounding the city of Cuzco. The chroniclers do not agree, however, on the location of these pillars, their number, or the precise way in which they were used.

The Incas also observed, but to a lesser extent, the passages of the Sun at the zenith. They did not count the years from a privileged date which would serve as a common reference, according to a practice in force in most civilizations.

As a result, they did not know their ages in years. The year was called *wata* (in Quechua) or *mara* (in Aymara). It was divided into twelve months or moons (*killa* in Quechua and *pacsi* in Aymara). In addition to months and days, there was also a division into 'weeks' of eight days. This was used to fix the frequency of markets, the services of the priests at the Temple of the Sun in Cuzco, and the services supplied to the emperor by the 'chosen women.'

8.3.2 Month Counting and Intercalation

The months, which all had the same number of days, were determined by observing the lunations. As a lunar year has about 354 days, a process of intercalation was necessary to keep the lunar year in line with the tropical year (or year of the seasons). This intercalation process consisted in adding one month approximately every three years. The procedure used for this intercalation is not specified by the chroniclers, Bernabé Cobo [1653] (1956) noting only that 'the remaining days of the year were integrated into the months themselves.' According to Sarmiento de Gamboa [1572] (1942, 2007), it was *Pachakuteq* who introduced this solar year with pseudo-lunar months. The names of the months of the year, as given by different chroniclers, are presented in Table 8.1.

8.4 The *Seq'es* of Cuzco

8.4.1 Bernabé Cobo and Other Chroniclers

The system of *seq'es* (or *ceques*) of Cuzco is known thanks to the writings of various Spanish chroniclers, including Cristóbal de Molina [ca. 1575] (1989), Juan Polo de Ondegardo [1585] (1916), José de Acosta [1590] (1954), Martín de Murúa [ca. 1615] (2008), Bernabé Cobo [1653] (1956), Cristóbal de Albornoz [ca. 1582] (1984), and Juan de Matienzo [1567] (1967). Brief biographies of these writers can be found in Sect. 3.3. Here, we shall mainly discuss the contributions of the Jesuit Bernabé Cobo, who wrote the *Historia de Nuevo Mundo* [1653] (1956). This work contains four chapters entitled *Relacion de las huacas*, devoted to the description of *wakas*, that is, sacred objects and places.

It seems that the information given by Bernabé Cobo was taken from another manuscript whose origin is not exactly known. According to the discussion by Bauer (1998), this anonymous manuscript would have been published between 1559 and 1572. The author himself would have based his text on information gleaned from the *khipus* masters who kept the records at the temples of Cuzco and the offerings made in such places. Polo de Ondegardo may have been the author of this original manuscript.

Table 8.1 Names of the months of the Inca calendar as given by the different chroniclers. [1] Juan Diez de Betanzos [1551] (1987). [2] Anonymous [ca. 1570] (1906). [3] Cristóbal de Molina [ca. 1575] (1989). [4] Polo of Ondegardo [1585] (1916). [5] Guáman Poma de Ayala [1615] (1936, 1980). [6] Gregorian equivalent. The year is believed to have begun at the December solstice, according to the writings of Juan Diez de Betanzos, Polo de Ondegardo, and Guáman Poma de Ayala. The spelling here is the one adopted by the chroniclers. After Bauer & Dearborn (1995)

[1]	[2]	[3]	[4]	[5]	[6]
Hatumpo Coquis	Hatumpocoy	Camayquilla	Camay	Capac Raymi Camay Quilla	January
Allapo Coiquis	Pachapocoy	Atunpucuy	Hatun Pucuy	Paucar Uaray Hatun Pocoy Quilla	February
Pacha Pocoiquis Ayriguaquis	Ayriuaquilla Haocaycusqui	Pachapucu Paucarguara	Pacha Pucuy Antihuaquiz	Pacha Pocuy Quilla Ynca Raymi Mamay Quilla	March April
Haucai Quos Quiquilla	Aymoayquilla	Ayriguay	Hatun Cusqu Raymoray	Atun Cusqui Aymoray Quilla	May
Hatun Quosquiquilla	Hatuncusqui	Hacicay Llusque	Aucay Cuzqui	Haucay Cusqui Quilla	June
Caguaquis	Chauaruay	Cauay	Chahua Huarquis	Chacra Conacuy Quilla	July
Carpayquis	Tarpuyquilla	Moronpassa Tarpuiquilla	Yapaquis	Chacra Yapuy Quilla Quilla	August
Satuaiquis Omarime Quis	Cituaquilla Chaupicusqui	Coyaraymi Omacrayma	Coya Raymi Homa Raymi Puchayquis	Coya Raymi Quilla Uma Raymi Quilla	September October
Cantaraiquis Pucoi Quillaraimequis	Raymiquilla Camayquilla	Ayarmaca Capac Raymi	Ayamarca Capacraymi	Aya Marcay Quilla Capac Ynti Raymi	November December

Much more recently, the *seq'e* system has been studied in detail by Zuidema (1964) and by Bauer (1998) in his work entitled *The Sacred Landscape of the Inca. The Cuzco Ceque System*.

8.4.2 The Radiating Structure of the *Seq'es*

At its peak in the sixteenth century, the Inca empire probably had a population between 12 and 15 million inhabitants. The city of Cuzco was the capital of this extensive territory until the arrival of the Spanish in 1532. In the center of the city was the *Qorikancha*, which was the epicenter of many rituals. Like most cities in the Andes, Cuzco was dotted with many *wakas*, where local communities worshiped with offerings and sacrifices in honor of deities and other idols. Unfortunately, many of these sacred places were ransacked and destroyed by the Spanish a few years after the conquest of the empire as part of their campaign to eradicate idolatry and destroy the native people's belief systems, so that they could forcibly introduce Christianity (Duviols, 1971).

In addition to structuring Cuzco into four districts, Bernabé Cobo [1653] (1956) mentions that the region surrounding Cuzco, within a radius of approximately 20 km, was divided up by abstract lines called *seq'es*, radiating out from the center of the city. Along each of these lines there were *wakas*—eight on average, but some had only three. The radial form of this system clearly indicates the importance of Cuzco, the Inca administrative capital and power center in *Tawantinsuyu*. Indeed, this structure testifies to the fact that power radiated out from this city.

After the introduction to his work, Cobo gives the name of each *seq'e*, each *waka*, and even the *ayllu* responsible for its maintenance. The quality of the information is not uniform. Sometimes it is imprecise and vague, but in certain cases it is enough to locate, without ambiguity, the described *waka*. On the whole, this text is a valuable source of information for reconstructing lost toponymy, locating forgotten places and buildings, and establishing the social geography of the ancient capital of the empire.

According to Juan Diez de Betanzos [1551] (1987), it was the *Sapa Inca Pachakuteq* who created the *seq'e* system. It has been suggested that the entire system of *seq'es* and *wakas* in fact represented the materialization in space of a calendar *khipu* (Williams, 2001).

The three neighborhoods *Chinchaysuyu*, *Antisuyu*, and *Qollasuyu* contained nine *seq'es* each, while *Kuntisuyu* had fifteen (and not fourteen as indicated by Cobo). In his text *Relacion de las huacas*, Cobo presents the *wakas* located along the different lines and indicates their respective distances from the Temple of the Sun. At the time of the arrival of the Spanish conquerors, there were at least

328 *wakas* (perhaps 400) distributed along the 41 or 42 *seq'es* emanating from the *Qorikancha*. *Chinchaysuyu* had at least 85 sacred places, *Antisuyu* 78, and *Qollasuyu* 85, while in *Kuntisuyu*, whose organization was more complex than that of the other districts, there were at least 80. The temples were divided into three categories characterized by the terms *qollana* (main), *payan* (secondary), and *kayao* (original), which seemed to reflect the degree of prestige of these holy places.

In the literature, it has generally been assumed that the *seq'es* were straight lines, but more detailed studies have shown that this assumption is not necessarily true. It seems established that the spatial divisions of the Cuzco valley specified by the *seq'es* were directly linked to the social organization of the capital and the responsibilities held by each clan regarding the various rituals. Officials from places of worship in the Cuzco region ensured that offerings in the various temples were made according to the planned rites and at the required times of the year. The upkeep and maintenance of these sacred places was carried out by the various lineages of the city.

The monuments marking out the *seq'es* made it possible to connect the *Qorikancha* to the mountain peaks, rocks, and remarkable features of the landscape that the Incas honored. Indeed, the mountain peaks were thought to be inhabited by the *apus*, powerful spirits before whom it was appropriate to bow with respect.

Considering the structure of the *seq'es*, we may say that the Incas genuinely reformed the conception of space. Thus, by connecting the whole Cusqueñian region with the *Qorikancha*, which was the most venerated place in *Tawantinsuyu*, they extended the limits of their city to the entire visible horizon. In this way, these lines constituted a kind of act of expansion, an assertion of their power and domination over the region. But these lines also served other purposes according to the designs of the Incas. They made it possible to receive the force which emanated from the peaks, the ravines, and the rocks. More precisely, these lines could serve as conduits providing energy and telluric force to this people of courageous warriors, accomplished builders, and extraordinary administrators. Likewise, the lines could work in the opposite direction, as channels of communication with the *apus* hidden in the heart of the mountains.

It has been suggested by some researchers (for a detailed discussion see Bauer & Dearborn, 1995) that certain lines of the *seq'e* system could have been used for observation of the heliacal rising or setting of certain stars. According to this hypothesis, these lines were drawn strictly along straight lines starting from the Temple of the Sun. Thus Zuidema (1982b) suggested that the sixth, eighth, and ninth *seq'es* of *Chinchaysuyu* corresponded to the setting of Vega, the Pleiades, and Betelgeuse, respectively, while the fifth *seq'e* of *Antisuyu* indicated the direction of the rising of the Pleiades and the first *seq'e* of *Kuntisuyu* the rising of β Centauri. These stars did indeed play an important role in Inca astronomy. However, detailed field studies have shown that the *wakas* on these *seq'es* were not arranged in perfectly straight lines, whence the suggested astronomical deductions look somewhat inappropriate.

8.4.3 On the Nature of the *Wakas*

The *wakas* of the Cuzco region were either of natural origin (mountains, springs, caves, etc.) or they were the result of constructions (houses, temples, canals, etc.). Some of these places became sacred in connection with the mythical history of the city of Cuzco and the Inca empire. Others bore a relationship with the calendar and indicated points on the horizon which marked special moments of the astronomical year.

According to Bauer (1998), we may group the 328 *wakas* of the Cuzco region into different categories, as summarized in Table 8.2. The largest categories are springs, rocks, and standing stones. Next, in decreasing numerical importance, are the mountains and hills, palaces and temples, squares, tombs, and ravines. All these sacred places were used to communicate with the supernatural world, mainly through offerings. Worshiping *wakas* was said to protect against illness, prevent violent death, help achieve victory in war, protect crops against bad weather, protect travelers, and many other things.

The importance of the offerings in the different temples varied depending on their size, but it seems established that, in the main temples, human sacrifices were carried out, particularly of young children. These took place especially at the time of the ritual of *Qhapaq Hucha* (the 'Great Gift'), a ceremonial

Table 8.2 Distribution of 328 *wakas* according to type. After Bauer (1998)

Nature of *waka*	Number	Percentage
Springs	96	29
Standing stones	95	29
Hills and passages	32	10
Palaces and temples	28	9
Fields, open places	28	9
Tombs	10	3
Ravines	7	2
Rocks, caves, trees, roads	16	5
Others	16	5

event during which visitors flocked to most of the temples of the empire (see Sect. 8.4.6). In many temples, offerings were either buried or burned. These could be fabrics that were cremated or precious objects (especially gold or silver) that were buried. Among the usual offerings were coca leaves, llamas, and marine shells, the most famous of which were spondyls, which could be crushed, ground into powder, or sculpted into figurines.

In addition to the temple of *Qorikancha*, the mountain of Huanacauri was particularly honored because the brother of the first mythical Inca, *Manqo Qhapaq*, was reputed to have turned into stone there. Other *wakas* were considered sacred as a result of their association with myths or legends, particularly those of the origin of the city and the empire. About fifteen *wakas*, in the form of stones, were linked to the war the Incas waged against the Chancas, an aggressive people living in the region of Ayacucho and Apurimac, who attacked the Incas during the reign of the eighth king *Wirakocha Inca*. According to legend, stones from the Cuzco region transformed into warriors and contributed to the victory of *Pachacuteq Inca*.

According to the *Relacion de la huacas* of Bernabé Cobo [1653] (1956), different temples were associated with successive sovereigns: the mummy of *Sinchi Roq'a* (the second Inca) was kept in a building called Acoyguaci, the temples of Sank'a Kancha and Tampukancha honored *Mayta Qhapaq* (the fourth *Sapa Inca*), the *wakas* of Guayllaurcaja and Taxanamaro were linked to *Wirakocha Inca* (the eighth sovereign), and Cugiguaman and Quinoapuquiu were associated with *Tupaq Inca Yupanki*. No fewer than five temples had their name linked to *Wayna Qhapaq* (the eleventh Inca).

It should be noted that some honored objects were no longer located in their original location, but had been transported to the Cuzco region. Finally, others were brought to the battlefields in the hope of influencing the outcome of battles involving the imperial armies.

8.4.4 *Wakas* and Inca Social Organization

According to Bernabé Cobo, the lineages of Cuzco were responsible for the tributes paid in the different *wakas* lying on the *seq'es* and were responsible for their maintenance. In particular, they had to check that sacrifices were carried out at the required times. There was therefore a relationship between the social organization of the capital and the maintenance of the temples. So, in this system, the collaboration of local lineages was achieved through the maintenance of a vast complex of sanctuaries, and at the same time this ensured the participation of the faithful in earthly acts of worship and respect for natural forces. This was ultimately a way of asserting and extending the status and power of members of the Inca elite.

The temple of *Qorikancha* was the site of complex rituals. Indeed, it was divided into several parts which corresponded to different *wakas*, attributed to several *suyus*. The names of queens (*qoyas*) also appeared in certain sacred places. At the death of the king, the male descendants of the deceased were responsible for the worship rendered to the deceased ancestor. According to the *Relacion de las huacas*, maintenance of the temples in the Cuzco region was ensured in part by the *panakas* of the ruling power, and in part by non-royal *ayllus*. According to the writings due to Cristóbal de Molina [ca.1575] (1989), there were ten royal and ten non-royal *panakas* in Cuzco. There were three in each of *Qollasuyu* and *Chinchaysuyu*, and two in each of *Antisuyu* and *Kuntisuyu*. The *panakas* linked to the first five sovereigns were traditionally associated with the *Urin* part of Cuzco and the next five with the *Hanan* part of the capital.

The relationship between the Cuzco *seq'e* system and the Inca calendar was studied by Aveni (1977). Noting that there were 41 *seq'es* comprising 328 *wakas*, we observe that $328 = 41 \times 8 = 27\ 1/3 \times 12$, where 8 is the length of the Inca week and 27 1/3 is the length of a sidereal lunar month. The 37 d missing from the tropical year correspond to the days between the heliacal setting of the Pleiades on 3 May and the heliacal rising of the same on 9 June, which coincides with a period of inactivity of the Incas. Following their heliacal rising, the duration of the presence of the Pleiades in the night sky increases with the length of the day at a time of year when the climate becomes increasingly warmer. The duration of the presence of these stars subsequently decreases with the length of the day when the climate becomes gradually colder.

Some authors have claimed that each day of the year corresponds to a *seq'e* and that many of them had an astronomical purpose. Thus, according to Cobo [1653] (1956), the *seq'es* could serve as sight lines on the horizon and be used to determine the June and December solstices as well as the sowing periods in the month of August. Zuidena (1995) has suggested that the *seq'es* were involved in determining the heliacal rising and setting of certain stars such as the Pleiades, the Southern Cross, α and β Centauri, and also the passage of the Sun at the zenith. They may also have helped with scheduling events.

8.4.5 Are There Any *Seq'es* Outside Cuzco?

One particularly important *seq'e* is the one setting out from the capital of the empire towards the site of Vilcanota, a place which today indicates the limit between the regions of Cuzco and Puno. The priests went on a pilgrimage along this *seq'e* at the time of the June solstice because the Sun was reputed to be born in this direction. Located 150 km from Cuzco, this temple in fact marked the boundary between Inca territory and that of the Qollas in the Lake Titicaca region. The details of the routes followed by the priests are even given by Cristóbal de Molina [ca. 1575] (1989), who indicates the list of sacred places visited on this occasion.

Systems comprising rectilinear lines coming from a central point in the manner of the system of *seq'es* of Cuzco undoubtedly existed in other places in the Andes, as discussed by Bauer (1998). Some have suggested this possibility, but without certainty, for Anta, Huanuco Pampa, and Inkawasi.

Some authors have attempted to establish an analogy between the Nazca lines and the system of *seq'es* of Cuzco. It was Toribio Mejia Xesspe (1896–1983), in 1940, who first used the term *seq'e* for the Nazca lines. The comparison is, however, unconvincing because major differences exist between the two sets of lines, quite apart from the fact that they date from very different periods. In the case of the *seq'es*, only one center exists (*Qorikancha*), while for the Nazca lines, Aveni (2000) identified around sixty centers from which around 750 lines emerge. The Nazca lines have clearly defined beginnings and ends, and extend over short distances along which there are few significant features, whereas in the case of the *seq'es*, many *wakas* were distributed along these lines.

8.4.6 The Ceremony of *Qhapaq Hucha*

The chronicler de Molina [1575] (1989) established a relationship between the ritual of *Qhapaq Hucha* (the 'Great Gift') and the *seq'e* system. This ceremony was not associated with a specific date during the year, but took place during major events such as the accession to the throne of a new *Sapa Inca* or during a large-scale natural disaster. According to other authors such as Cieza de León [1553] (1984), it could be repeated annually.

When the *Qhapaq Hucha* ritual took place, offerings were collected in the different villages. They were made up of precious fabrics, spondyls, jewelry, livestock, but also children around ten years old. During the ceremony, objects were burned and the hearts of llamas and human beings were torn out to be offered to the deities. These victims were adorned with beautiful finery before being sacrificed to the deities after grandiose ceremonies. They were strangled or drowned after having absorbed anesthetic substances but, as the ultimate reward, they were destined to be venerated after their death. Human sacrifices to the mountain gods are attested by the mummies of children and young girls whose well-preserved bodies have been found in the Andean glaciers. These sacrificial practices had their origins in more ancient cultural contexts, as evidenced by certain Mochica terracottas.

The priests then traveled to the most important temples of the empire to transmit part of the offerings. They walked in a straight line over hill and valley from the *Qorikancha* to reach the various sacred places and transport part of the sacrificial material there. The *caciques* who had participated in the *Qhapaq Hucha* ceremonies in Cuzco then returned to their villages after receiving gifts from the *Sapa Inca* (textiles, goldwork, servants, secondary women, etc.).

9

Worship of the Sun

9.1 Introduction

Pachakuteq created the system of *wakas* and *seq'es* emanating in all directions from the *Qorikancha*. He had Sun temples built in many places across the empire so that everyone could honor the Sun (Cobo, [1653] 1956). In addition to the particularly venerated *Qorikancha* in Cuzco, the site of Sunturhuasi and the *usnu* of Hanan Hawkaypata square were also among the privileged places where one could honor the day star. Many other *wakas* were frequented, but most were destroyed by the Spanish conquerors in their attempt to eliminate idolatry, as they put it at the time. Many of the surviving *wakas* were made of rock, most of which were carved. A detailed examination of these shows that a large number of these sites were characterized by a favored astronomical orientation (Gullberg, 2009, 2020). But a certain number of them were not, although identifying such preferred directions is not always obvious (Urton, 1978).

9.2 An Omnipresent Belief: The Cult of the Sun

For the people of the Andes, and the Incas in particular, astronomy cannot be dissociated from mythology, religion, and agriculture. To understand it better, we must understand the way of life of this society, placing ourselves at the heart of their culture and civilization. Inca astronomy finds its roots in the older Andean civilizations because the Incas knew how to bring together and integrate contributions from the cultures that preceded them. Therefore, to better understand this astronomy, we must also consider the contributions of previous civilizations (see Chap. 4).

We know that, according to the Incas, the world was created by *Wiraqocha* near Lake Titicaca. This god was the father of the Sun and the Moon. Both male and female, he was therefore the founder of the patrilineal and matrilineal lineages. He created the Sun, Moon, and stars while human beings were formed from rocks or boulders. He gave life to the stones when they appeared from the caves where they originated. So, the Incas proclaimed themselves the sons of the Sun. The cult of the day star constituted the main religion of the empire. *Pachakuteq* introduced the idea that he was the son of the Sun. The emperor was the main figure of the religion because he was set up as the intermediary between the Sun and the people (Bauer & Stanish, 2001). This made it possible to establish a state religion and thereby consolidate the power of the reigning sovereign. The queen, the Qoya, was compared to the Moon. She was the sister and wife of the sovereign (Baudoin, 2002).

The Inca pantheon was very extensive (see Sect. 6.4). According to a widespread myth (see Sect. 5.1), *Manqo Qhapaq* left the surroundings of Lake Titicaca with his brothers and sisters for a migration which would take them to a cave south of Cuzco (Hemming & Ranney, 1982). Caves played a vital role in Inca mythology because they establish a relationship between the terrestrial world and the underworld. According to legend, the islands of the Sun (Isla del Sol) and the Moon (Isla de la Luna) on Lake Titicaca were the points of origin of the stars of day and night, and hence of the whole Inca nation. *Pachaquteq* built temples there and these islands became popular places of pilgrimage. It was believed that the Sun first appeared from a rock called Titicala in the north of the Island of the Sun and ceremonies were held there at the June and December solstices.

We have seen that the Incas distinguished three worlds in their cosmology: the underworld, the terrestrial world, and the upper world. In many places in the empire, stairs symbolized the transition, the passage between these three universes (Urton, 1981a).

The Incas revered mountains, caves, springs, and rivers which they believed to be invested with superior power. Mountains were considered the residence of the gods, or indeed as embodying powerful deities (Reinhard, 1985). Many rocks, whether sculpted or not, reflected manifestations of the sacred and materialized places where the divine was honored.

9.3 A Very Ancient Solstice Alignment

A common expression at many pre-Inca sites is the presence of ceremonial platforms (*usnus*) and temples (*wakas*) comprising staircases with axes of symmetry oriented toward the rising or setting of the Sun at the time of the solstices. This type of ritual architecture dates back as far as 3100 BCE (Malville, 2015). These sites gave the public the opportunity to participate in ritual ceremonies and to witness the rising and setting of the day star at these special moments.

The presence of cosmological motifs in Andean architecture dates back to the Archaic Period (3100–1800 BCE), if we are to believe Haas & Creamer (2006, 2012) and Shady Solis (2006), and would be characteristic of the platform mounds found in the Norte Chico region (approximately 200 km north of Lima), particularly in Caral. This practice would have continued throughout the Initial Period (1800–1000 BCE) in the Casma Valley, to be found in Chavín de Huántar during the Early Horizon (1000–200 BCE). Many of these platforms were oriented towards sunrise at the time of the June solstice and sunset at the time of the December solstice. On the larger mounds, stairs rose from circular plazas below and evoked the symbolism of a passage through the three worlds. The platforms materialized artificial mountains and anticipated the Inca *usnus*, used for ritual representations during which different kinds of liquid (water, beer, blood) were offered to the *apus*, the gods of the mountain (Staller, 2008).

The Great Pyramid of Caral has a circular courtyard below, which is accessed by stairs. The face of the pyramid is parallel to a line with azimuths 114°/294° corresponding to sunrise at the winter solstice and sunset at the June solstice. Other ancient structures have similar orientations (Malville, 2015). They are found at Sechin Bajo in the Casma Valley (dated to 3500–3000 BCE) and at Buena Vista in the Chillon Valley (2200–1750 BCE). A U-shaped structure in Chavín de Huántar (1000–200 BCE) is also oriented towards the sunrise on the December solstice (Rick, 2008).

As reported by Malville (2015), it is remarkable to note that the main axis of orientation of ten sites out of the thirteen in the Casma valley corresponds to the direction of sunset at the June solstice or sunrise at the December solstice. This clearly indicates that these directions were deliberately chosen. Residential structures in San Diego (Ancient Horizon) also present an orientation linked to the directions characterizing the winter and summer solstices.

The thirteen characteristic towers of Chankillo (Casma Valley) have attracted the attention of many archaeo-astronomers (see, for example, Ghezzi & Ruggles (2007, 2011)). According to carbon-14 dating, these towers could date back to the period 400–100 BCE. Several square areas at this site would have played an important role during official ceremonies. They are oriented in such a way that the celebrants located in these squares could observe the sunrises and sunsets during the solstices.

Two other ceremonial centers also feature structures whose orientation is linked to the positions of the Sun at certain times of the astronomical year. One is Tiwanaku, not far from Lake Titicaca, and the other is the sanctuary of the Island of the Sun (Isla del Sol). In Tiwanaku, the complex called Kalasasaya is oriented along an east–west axis. It contains ten andesite monoliths, each four metres tall, which have been interpreted as part of a calendar device (Zuidema, 2009). Seen from the center of the platform, the southernmost monolith corresponds to the position of sunrise at the December solstice, while the northernmost one corresponds to the position of sunset at the June solstice. On the site of the Island of the Sun, we find the sacred rock from which the original Sun emerged according to legend. Behind it, two pillars located on the horizon indicate the position of the sunset during the June solstice.

9.4 A Sacred Territory

Wakas and *usnus* are omnipresent elements of Andean cosmology that have been found since the civilization of Chavín de Huántar. The Lanzón of Chavín is a well-known example of a *waka* whose origins go very far back in time. A transcendent element of this cosmology is the shamanic transition between the three worlds, often achieved through symbolic staircases. This was particularly the case at Chavín, Chankillo, and Chinchero, but also at Machu Picchu (Malville, 2010). One of the primordial goals of the Andean religion and associated rites was to preserve a balanced and harmonious universe. This is the reason why offerings were presented at many sacred places. The *wakas* possessed a supernatural power and also a power of animation for people, animals, and also domestic herds (Malville, 2010).

The study of *wakas* indicates that these were considered to be places with extraordinary power derived from the wisdom of the ancestors. They could function as oracles and, in some cases, they could be destroyed by enemies. The personality specific to these places erased the distinction between the animate and the inanimate world (Bray, 2009). The 328 sacred places located near Cuzco, spread over 41 *seq'es*, could in some way be considered as wielding a power of protection over the Inca empire. The pillars marking the position of sunrises and sunsets on the horizon were important to clarify the progress of the tropical year. During the *Inti Raymi* festivities at the June solstice, the *Sapa Inca* and his relatives contemplated the sunrise from Hawkaypata square, while the common people observed it from Kusipata square, west of the Watanay River.

Water was an animating force specific to Andean cultures. This energy transmitting agent was known as a *kamaq* (see Sect. 6.2). Thus the dark constellation of the Llama in the sky was the *kamaq* of llamas, responsible for giving these animals the vitality they need to flourish on Earth. The Pleiades (Qolca), known as the storehouse or granary, was particularly revered because this asterism was considered the supreme *kamaq* from which came all the energy needed to give life to animals (Cobo, [1653] (1956)).

9.5 Rocks and Astronomical Alignments

Most of the *wakas* that survived the campaign to eradicate idolatry by the Spanish conquerors were constituted by rocks. Gullberg (2009, 2020) carried out a detailed study of these *wakas*. Thirty-one different sites were studied: nineteen were in the regions of Cuzco, Tipon, and Saihuite, nine in the Sacred Valley, and three in the Machu Picchu region. These sites can be associated with characteristic astronomical directions (solstice or equinox). They are grouped in Table 9.1.

Among these places, *intiwatanas* are found at the site of Machu Picchu, near the Urubamba River, at Tipon, and at P'isaq (Hemming & Ranney, 1982). Gullberg (2009, 2020) notes that the Sacred Square of Machu Picchu, the *intiwatana* of the Urubamba River, and the Temple of the Sun in Llactapata lie approximately on an axis defined by the sunrise at the June solstice and sunset at the December solstice. These three places may have constituted a local *seq'e*.

The semi-circular structure of Torreón includes two openings allowing the observation of the sunrise during the June solstice and the heliacal rising of the Pleiades (Dearborn & White, 1989). The Torreón could also have been used to observe the passage of the Sun at the zenith. The Sacred Square and the Torreón are aligned along an axis corresponding to the sunrise during the June solstice and the sunset during the December solstice.

In the eastern part of the urban area of Machu Picchu is the Intimachay. This structure may have been built to observe the sunrise at the December solstice, with light entering the tunnel for approximately ten days before and after the solstice (Dearborn, Schreiber & White, 1987; Dearborn & White, 1989). It could also be an astronomical observatory built for calendar purposes in relation to the festival of *Qhapaq Raymi* (Ziolkowski, Kosciuk & Astete Victoria, 2013).

Table 9.1 *Wakas* characterized by astronomically aligned rocks near Machu Picchu, in the Sacred Valley, and in the Cuzco region. The sites mentioned in the table are characterized by the presence of an astronomically aligned rock. In the case of symbols in brackets, the structure concerned is not a rock. According to Gullberg (2009, 2020)

Site	LSSJ	CSSJ	LSSD	CSSD	ESE	CSS	LSZ	LSAZ
Machu Picchu								
Machu Picchu	x	x	x	x	x		x	x
Intiwatana of the river Urubamba					(x)		(x)	
Llactapata	(x)							
Sacred valley								
Cerro Unoraqui	(x)							
Chinchero	x			x	x	x		
Choquequilla			x					
Moray	(x)						(x)	
P'isaq	x		x					
Q'espiwanka	x	x	x					
Ollantaytambo	x	x	x				x	
Cuzco region								
Waka 44	x	x	x	x	x	x	x	
Q'enqo Chico			x					
Q'enqo Grande	x		x		x	x		
Kusicallanca			(x)					
Kusilluchayoc		x						
Lacco	x						x	
Lanlakuyok					x			
Mollaguanca	x			x				
Puca Pucara					(x)	(x)		
Rumiwasi Alto			(x)					
Rumiwasi Bajo			(x)					
Saqsaywaman							(x)	(x)
Saihuite	x							
Tambomachay			x					
Tipon		x						

LSSJ: sunrise at the June solstice
CSSJ: sunset at the June solstice
LSSD; sunrise at the December solstice
CSSD: sunset at the December solstice
ESE: sunrise at the equinox
CSS: sunset at the solstice
LSZ: passage of the Sun at the zenith
LSAZ: passage of the Sun at the anti-zenith (nadir)

In a building in the urban area, there are two small circular sculptures called 'mortars.' The role played by these small pools, which may be filled with water, is unclear. They could have been used for indirect observation of the Sun or Moon, or to reflect sunlight entering through a window during the equinoxes. An observation of the Southern Cross cannot be excluded, this constellation being honored by the Incas because, if we are to believe Zuidema (1982a; 1982b), the fourteenth *seq'e* of *Kuntisuyu* was aligned with the rising of this constellation.

On the northwest face of Huayna Picchu is the Temple of the Moon, a site that contains two caves and may have housed mummies. Three niches on the site are approximately aligned with the sunset during the June solstice, and they can also be illuminated by the Moon.

The Llactapata site is about five kilometers from Machu Picchu. It has been carefully excavated quite recently (Malville, Thomson & Ziegler, 2006). The Temple of the Sun has an orientation similar to that of the *Qorikancha* in Cuzco. A stone-lined canal in front of a main gate and crossing the Sacred Square is oriented in the direction of the sunrise at the June solstice and in the direction of the heliacal rising of the Pleiades. An adjacent thirty-three meter long corridor is aligned in these same directions (Gullberg, 2009, 2020). We know that the direction corresponding to the heliacal rising of the Pleiades was also favored in the case of the *Qorikancha* (Zuidema, 1982a, 1982b).

Observation of the Pleiades was of great importance to the Incas for planning the harvests. When the stars of this constellation became very bright, it was appropriate to rejoice because it meant lots of rain and good harvests in the following months. On the other hand, if the stars appeared duller, people should fear a period of drought (Orlove, Chiang & Cane, 2000). These simple observations made it possible to anticipate the effects of *El Niño*. Indeed, similar observations are still made today by Andean farmers.

9.6 Festivals and the Rhythms of Time

A ritual calendar was established by *Pachaquteq* and was determined by the positions of sunrise and sunset on the horizon. The religious festivals and celebrations organized in relation to the agrarian year had a social and political vocation. Festivals thus marked important times of the year for planting, sowing, or harvesting, and more generally, the seasons. They also strengthened the authority of the *Sapa Inca* and the ruling elites.

Among these festivals were Pacha Pucuy at the March equinox, *Inti Raymi* at the June solstice, *Qoya Raymi* at the September equinox, and *Qhapaq Raymi* at the December solstice. *Qhapaq Raymi*, the summer festival, marked the time when crops would begin to germinate and the beginning of a new season. They also reflected the ritual in which adolescents would make the transition to adulthood (Hemming 1971). *Inti Raymi* was the most important festival of the year. It attracted many pilgrims to Cuzco during the June solstice. These were elaborate festivities that lasted several days and during which numerous offerings were made to the gods (Zuidema, 1986; Bauer & Stanish, 2001).

9.7 Agriculture and Its Relationship with Observations of the Moon and Sun

The efficiency of Inca agriculture was such that many citizens could devote themselves to military or construction tasks. Each family had plots of land located in various ecological conditions. A wide variety of crops were grown by each *ayllu*. In the highlands, there were cereals, such as quinoa, and countless varieties of potatoes. Dried potatoes (*chuño*) were easy to store. Corn was frequently consumed in soups and also used to make beer (*chicha*), which flowed in abundance during festivals (Fig. 9.1).

The coastal oases produced many foods: squash, beans, peppers, avocados, sweet potatoes, cassava, and peanuts, in particular. Coca was reserved for the dominant elite and served as a stimulant to alleviate hunger or fatigue. It was also used as an offering to the deities.

Countless terraces, arranged on the mountainside, and irrigated using a vast network of canals, made it possible to cultivate the smallest plot of land while controlling erosion. The Incas adopted the concept of terraced cultivation from the Waris, who themselves took it from their ancestors (Wright & Valencia, 2000). Floodable or marshy areas were also exploited when suitable for agriculture. Sources, fountains, and irrigation canals played a particularly important role in Peru, because the Incas were masters in the art of irrigating land. Fountains were important for both practical and ceremonial reasons. Good examples can be found in Ollantaytambo, Tipon, or Machu Picchu.

The crop calendar (planting, sowing, harvesting) was determined by observation of the Sun and the Moon. The presence of pillars placed on the heights of Cuzco were used to establish the most favorable periods for planting corn (Urton, 1978b, 1981a), the ideal season for cultivating this herbaceous plant extending from October to May.

Fig. 9.1 Corn was abundantly consumed in the Inca civilization. We see here the grain crushing operation, the flour being used to make pancakes. Author's photograph

9.8 Horizon Astronomy and Archaeoastronomy

Sunrises and Sunsets

Detailed observation of the constructions of ancient civilizations often indicates that the buildings were not arranged randomly but were characterized by a privileged orientation linked to observation of the Sun, the Moon, the planets, and certain stars such as the Pleiades. It is one of the goals of archaeoastronomy to identify such alignments and to determine why they were favored (Zawaski & Malville, 2007–2008). As we know, the Inca civilization attached great importance to what is referred to as 'horizon astronomy', that is, astronomy carried out with the naked eye. This field relating to the search for astronomical alignments has developed considerably over the last few decades (see, for example, Aveni, 1980, 2003; Bauer & Dearborn, 1995).

At mid-latitudes, celestial bodies rise in the east, cross the sky in an arc, and set in the west. Seen from the southern hemisphere when looking north, these bodies move from right to left and, in the northern hemisphere when looking south, from left to right. The altitude of the arc in the sky varies during the year and depends on the latitude of the observer and the position of the Earth on its annual trip round the Sun. This annual variation is manifested by a shift in the points of sunrise and sunset on the horizon.

The Inca capital, Cuzco, is not far from the equator. More precisely, it is located at a southern latitude of $13°\ 29'\ 1''$. In this city, for an observer looking north, the Sun will rise in the east (right side), follow an arc across the sky toward the north, and set in the west (left side). During the June solstice, the Sun will rise in the northeast, follow a small arc in the sky, and set on the northwest side. During the December solstice, the Sun will rise in the southeast, follow a longer arc, and set in the southwest. The observation of the different positions of sunrises and sunsets on the horizon during the year was the subject of much attention on the part of the Incas. They used these observations to develop a calendar system to determine the dates for planting, sowing, and harvesting and to set the dates of festivities during the tropical year. Moreover, on the heights of Cuzco, they built a system of pillars allowing them to establish the privileged moments of the astronomical year with greater accuracy.

The times when the Sun, in the ecliptic plane, crosses the celestial equator correspond to the equinoxes. They occur around 21 March and 21 September and correspond to times of the year when days and nights have equal lengths. The times when the Sun reaches the northernmost and southernmost points of the ecliptic (corresponding to $23.44°$ north and south) are the solstices. They occur around 21 June and 21 December. For careful observers, the solstices are quite easy to observe and correspond to the period during which the position of the sunrise on the horizon more or less stops moving for a few days.

The cardinal directions would undoubtedly have been determined by the ancients by appealing to a gnomon. By watching the shadow cast by a gnomon during the day, we use the fact that the shortest shadow produced corresponds to the Sun being in the south. North is then in the opposite direction, while east and west correspond to the perpendicular directions (Evans, 1998).

The Passage of the Sun at the Zenith

It is well established that the observation of the positions of the Sun at the solstices and equinoxes has always been of major importance and that it has often been used to determine the orientation of buildings and monuments. In the region of the Earth's surface between the tropics of Cancer and Capricorn, another astronomical phenomenon occurs which has attracted less attention: it is the passage of the Sun at the zenith (or even at the nadir, the point located opposite the zenith in relation to the center of the Earth, but which itself is unobservable).

The zenith corresponds to the point on the celestial sphere located vertically above the observer. It is only in the region of the Earth's surface bounded by the tropics of Cancer and Capricorn that the Sun can be observed at its zenith. On the line of the tropics, only one zenithal passage of the Sun is observed each year, and it coincides with one of the solstices. On the tropic of Cancer, it is the summer solstice, while on the tropic of Capricorn, it is the winter solstice. At the equator, the passage at the zenith coincides with the equinoxes.

The passage of the Sun at the zenith, which corresponds to the highest point in the sky, depends on the latitude of the observer. This passage is easily observed using a gnomon, because at this moment no shadow is cast on the ground at noon. In the case of a deep well containing water, it is possible to see the Sun reflected by the water in the well at noon. It was this process that allowed Eratosthenes, in Greek Antiquity, to determine the circumference of the Earth. At the time of the summer solstice, he observed the reflection of the Sun at noon by the water in a well in Syene, then the shadow cast by a gnomon at the same time, but located in Alexandria.

Although it is well established that many ancient peoples throughout the world considered the astronomical moments corresponding to the sunrises and sunsets during the solstices and equinoxes to be very important, to the point of orienting the architecture of buildings and monuments accordingly, the architecture of other constructions seems to have been influenced by the passage of the Sun at the zenith.

In the vicinity of Cuzco, the Sun passes through the zenith twice a year, around 13 February and around 30 October (Zuidema, 1981). As mentioned above, for points located in the tropics at a latitude of 23.44°, the Sun passes vertically to that location during the associated solstice. Zuidema (1981) suggested that the Incas observed the passage of the Sun at the anti-zenith (the point located at the nadir, that is to say, opposite the zenith). At the latitude of Cuzco, the passage of the Sun at the anti-zenith takes place around 26 April and around 18 August. The argument is based on the fact that the festivals chosen by the Incas to celebrate the planting and harvest of corn took place approximately on these dates. However, we cannot find any convincing argument to clearly establish that the passage of the Sun at the anti-zenith was really 'observed' by the Incas.

9.9 Astronomical Observations at the *Qorikancha*

The *Qorikancha* was, as we have seen, the most important temple in Cuzco. It was a place filled with wealth, where the most sacred effigies of the empire were worshiped, including the *P'unchaw*, the golden representation of the Sun. From the period of the Spanish conquest, we have some information, unfortunately not very precise, indicating the orientation chosen for certain elements of the temple, including the *P'unchaw*. If we are to believe the chronicles of Pedro Cieza de León [1553] (1984), it was brightly illuminated during sunrise and sunset at certain times of the year. It should be noted that the Temple of the Sun was destroyed by earthquakes in 1650 and 1950.

From the ruins of the temple as they remain today, certain authors, including Zuidema (1977, 1981) and Aveni (1981), have formulated hypotheses about its astronomical vocation and deduced that the observations carried out there could have been used to establish a 328-day calendar. According to Zuidema (1981, 2010), the Incas carefully observed different times of the tropical year, namely:

- the solstices on 21 June and 21 December;
- the equinoxes on 21 March and 21 September;
- the passage of the Sun at the zenith which, in Cuzco, takes place around 13 February and 30 October;
- the passage of the Sun at the nadir (or at the anti-zenith) which, in Cuzco, occurs around 26 April and 18 August, this observation obviously not being direct;
- the sunrise when the Sun's rays pass down the main corridor of the *Qorikancha*, i.e., 25 May and 18 July, these dates contributing to the calendar organization of the year (Zuidema, 1981, 2010).

It seems obvious that certain observations could have been made from the *Qorikancha*, notably the passage of the Sun at the zenith. But researchers have suggested that other kinds of observation could also have been made from this temple, such as the sunrise during the June solstice and the heliacal rising of the Pleiades (Zuidema, 1982a, 1982b), or the sunset during the December solstice (Bauer & Dearborn, 1995).

Zuidema (1981) and Aveni (1981) postulated that the system of 41 *seq'es* centered on the *Qorikancha* played a particular role during the observations, and that certain *wakas* indicated privileged directions for solar observations. This is how the sunset during the December solstice could be observed from

Qorikancha, by looking towards the *sukanka* located on the hill of Chinchicalla. The eastern wall of the building was characterized by an azimuth of 66° 44′, which made it possible to observe the sunrise around 25 May (Gregorian calendar) and 18 July, and would thus have given the first of these dates great importance. This last alignment would also have favored the observation of the heliacal rising of the Pleiades, that is to say, their appearance in the sky about an hour before sunrise around 6–9 June (Gregorian calendar), while they would have been invisible since May. It is from these two dates of 25 May and 18 July that Zuidema (1982a, 1982b, 2014) put forward the hypothesis of the use of a *khipu*-type calendar comprising 328 days, based on an equal number of *wakas*.

Zuidema's hypotheses were subsequently criticized by Bauer & Dearborn (1995) and then by Ziolkowski & Kosciuk (2018). The latter in particular measured an azimuth of 67° 06′ (instead of 66° 44′), which corresponds to the observation of sunrises on 23 May and 19 July, with an interval of 57–58 days, and no longer 55 days/nights, a duration which no longer corresponds to a double lunar month ($2 \times 27.3 \approx 55$). The deductions of Zuidema (1981), according to these authors, should be revised accordingly.

9.10 Astronomical Observation Instruments and Sites

The calendar used by the Incas for agricultural and political purposes required detailed observation of celestial bodies like the Sun, the Moon, and also the Pleiades (Aveni, 1981; Zawasti, 2007). Astronomical knowledge played an important role during the expansion of the empire because it provided the Inca elites, and particularly the emperor, with the means to dominate the people through the control of rituals and the establishment of a state cosmology (Bauer & Dearborn, 1995).

Certain observation systems are mentioned in Spanish chronicles. This is the case of the gnomon used in particular to determine the passage of the Sun at the zenith. Some buildings were also oriented in directions favoring astronomical observations. We also saw that pillars located on the heights of Cuzco could be used to determine the passage of the Sun at the solstices.

During the transits of celestial bodies, two different categories of instrument could in fact be used. To observe the transit of the Sun at certain times of the astronomical year for religious and ceremonial reasons, the *sukankas* on Mount Picchu to the west of Cuzco were used. An observation instrument used for the same purpose consisted of a corridor leading to the main square of the

Ingapirca site in Ecuador. These 'observatories' did not particularly set out to achieve astronomical precision, but rather had the stated aim of producing visual effects that would be accessible to a large number of spectators during official ceremonies.

The Inca world also had observation sites which we could describe as astronomical observatories, but which were accessible only to a limited number of priest-astronomers who went there at selected times of the year. The Intimachay site at Machu Picchu is an example. Another is the site of Inkaraquay-El Mirador, also located in the Machu Picchu National Park.

Intimachay is a natural cave located on the eastern terraces of Machu Picchu. It was modified to serve as an observatory for astronomical purposes. The uniqueness of the structure and the precision of the alignments made during its construction attest that this is not the effect of chance. This site was described by Dearborn, Schreiber & White (1987) and, more recently, by Ziólkowski, Kościuk & Astete Victoria (2013) who provided many details. It was particularly suitable for observing the December solstice. The site had two windows, one on the north side and the other on the east side, which overlooked tunnels used for astronomical observations. It seems established that the window on the east side was used to observe the sunrise during the December solstice. Given the location of the windows, other phenomena could be observed, such as the sunrise during the equinoxes or the sunrise during the June solstice. Lunar observations cannot be ruled out either.

Inkaraqay-El Mirador is a small structure located on the northern slope of Huayna Picchu. This structure has been described in detail by Astete Victoria et al. (2016–2017). They note that it was equipped with niches and openings that were manufactured in a particularly careful manner to allow astronomical observations to be carried out. From this observatory, looking through the opening located to the north, one can observe the sunrise above Yanantin mountain during the June solstice. When the sunlight was too blinding, the aperture could have been used for observations of the shadow cast on a rear wall using a gnomon. The path of the Pleiades during their heliacal rising could also have been observed through these openings. The observation of this constellation was very important for the Incas, as we know, because it could be used to determine the time of the harvests, of great importance in the agrarian year.

It is more difficult to determine whether these apertures were also used to observe other stars in the sky, because little or no information is available on this subject. Bright stars such as Arcturus (α Boo) or Hamal (α Ari) could have been among such stars, if we are to believe Astete Victoria et al. (2016–2017), but this is far from being established. On the other hand, it would have been possible to observe the planet Venus when it reached its maximum northern declination, and this makes it a more serious candidate when we know the interest shown by the Incas in this planet.

10

Astronomy in the Andes

10.1 Hesiod and the Pleiades

The Andean peoples lived in a world that they considered to be animated by natural but also supernatural forces. These forces were believed to be present in both terrestrial and celestial objects. Given the importance it played in regulating the agrarian cycles of these peoples, the Sun was considered a supernatural power, and even before the advent of the Inca empire, this star was particularly honored on the islands of Lake Titicaca. Andean astronomy has long attracted the attention of researchers (see, for example, du Gourq 1893; Antonialdi 1942), but has aroused renewed interest since the end of the twentieth century (see, in particular, Aveni 1980; Krupp 1979, 1983; Urton 1981a; Bauer & Dearborn 1995; Gullberg 2009, 2020).

As we have seen, astronomical observations, in particular lunar and solar observations, played a major role in the establishment of the Inca calendar. These people used a rather simple calendar established from the observation of sunsets and sunrises on the heights of Cuzco. Solar observations were used to establish the dates of the December and June solstices in relation to the times of sowing and harvesting, and were the subject of major celebrations like *Qhapaq Raymi* and *Inti Raymi*. Observations of the equinoxes and the passage of the Sun through the zenith were also carefully considered by the Incas, but these events are less often commented in the available texts.

In his *Works and Days*, written in dactylic hexameters towards the end of the eighth century BCE, the Greek poet Hesiod sets out the list of agricultural tasks and maritime and social duties that must be carried out at different times of the year in order to live in harmony with the universe. Human activities must be synchronized with the rhythms and cycles occurring in the physical universe, that is, in the terrestrial and celestial worlds, for society to function effectively. Thus, he writes in his poem[1]: "Dedicate yourself with pleasure to useful works, so that your barns are filled with the fruits gathered during the auspicious season. It is work that multiplies herds and increases opulence. By working, you will be much dearer to gods and mortals because the idle are odious to them."

Regarding the harvest, he further advises: "Begin the harvest when the Pleiades, daughters of Atlas, rise in the heavens, and the plowing when they disappear; they remain hidden forty days and forty nights, and show themselves again when the year is over, at the time when the iron blades are sharpened. This is the general law of the countryside for settlers who live on the seashore or who, far from this stormy sea, cultivate fertile soil in the gorges of the deep valleys." Later in his text, he writes again about navigation: "If the desire for perilous navigation has taken possession of your soul, fear the time when the Pleiades, fleeing the impetuous Orion, plunge into the dark Ocean; then the breath of all the winds is unleashed; do not expose your ships to the fury of the dark sea."

Thus, already at the time of Hesiod, the observation of the Pleiades played an important role in planning agricultural work. It is therefore surprising and quite remarkable that the reappearance of the Pleiades in the sky, which marked the beginning of a new agrarian cycle among the Incas, was also the subject of particular attention on the part of this people. Thus, more than two thousand years apart in time and thousands of kilometres apart, two peoples with no common history realized the importance of the same celestial phenomena as a guide for their daily lives.

10.2 The Sources of Inca Astronomy

Regarding astronomy in the Andes in the pre-Columbian era, and during the Inca empire in particular, there are several sources. Some information comes from excavations carried out, not only around Cuzco, but also at other sites such as Incahuasi or Ingapirca in Ecuador. This is also the case for the Ayacucho region where the Wari civilization developed. The site of Machu Picchu and the region of Lake Titicaca have also been the subject of in-depth archaeological research. The location of the religious sites involved in the *seq'e* system near Cuzco has generated much debate and many studies. The identification of

---

[1] Translated by the author from the French translation at the website http://remacle.org/bloodwolf/poetes/falc/hesiode/travaux.htm.

wakas, particularly those located in the hills, is not easy and has required detailed investigations.

The astronomy practiced by the Incas is not known to us through works dealing specifically with this discipline, but through the writings of numerous Peruvian or Spanish chroniclers, dating for the most part from the end of the sixteenth century and the beginning of the seventeenth century. These writings relate to the history, way of life, and traditions of the Andean peoples. They are written in Quechua and Spanish, and some of them deal with subjects far from astronomy, such as the eradication of certain practices that the Spanish conquerors equated with idolatry. A list containing the names of the main chroniclers referred to in this chapter is given in Chap. 3 (see also the references at the end of the volume).

Other sources are the dictionaries of Quechua and Aymara published by the Spanish from the beginning of the conquest of the empire. All these works contain a large amount of information concerning rituals, festivals, calendars, and observations of the sky.

10.3 Agrarian Cycles and Astronomy

At the altitude of Cuzco, around 3400 m, we reach the limit for the cultivation of corn. In the mountains surrounding this town, wheat, potatoes, oca, ulluco, quinoa, and other high altitude crops were grown using the fallow system. The agricultural calendar had to take into account the durations of the different cycles, from corn, whose growth requires many months, to certain tubers with a shorter growing period.

In addition to the circadian rhythm associated with the rotation of the Earth around its own axis, the Spanish chronicles, suggest that the Incas used an eight-day agrarian week before the Spanish conquest (Zuidema 1977). Indeed, this is still in force in certain communities on the coast of Peru, although most Peruvians have now adopted the rhythm of the seven-day week of the Gregorian calendar. Synodic and sidereal lunar periods, and solar periods of approximately thirty days defined by the observation of the solstices, the equinoxes, and the passages of the Sun at the zenith and nadir established a correspondence between astronomical cycles, religious rituals, agricultural work, and political activity.

The agrarian cycles in the Cuzco region were not only defined by the periods for planting corn, potatoes, and cereals, but also by the tasks of irrigation, hoeing, and weeding which had to be dealt with at certain times of the year depending on the phases of plant growth. The sequence of work described by Urton (1981a) in the Andean community of Misminay, near Cuzco, is undoubtedly close to that followed by the Incas in the sixteenth and seventeenth centuries. The first plantings, those of early potatoes and wheat, take place in July. This is followed in August and September by the planting of high altitude tubers such as oca and ulluco, then corn and quinoa. Peas are planted in October and later potatoes from November to December. Festivities mark the planting of corn in September. They are accompanied by sacrifices and abundant consumption of *chicha*. One of the tasks carried out between November and February is hoeing, which consists in covering the base of the corn plants with soil from the furrows using a hoe (Fig. 10.1). The first operation of this type is carried out in November and the second in January. From mid-February, the early potato cycle ends. Other crops (quinoa, wheat, late potatoes, etc.) are harvested from the beginning of May until mid-June, a month during which the days can be cold and also dusty due to the absence of rain.

Most of these agrarian activities last about a month or even a month and a half. According to Urton (1981a), in the community of Misminay, Andean farmers consider it important to plant potatoes, ocas, and ullucos during the

Fig. 10.1 The Incas did not use beasts of burden. The tools used for working the land were rudimentary. Author's photograph

waning moon phases, while it is appropriate to sow surface crops such as corn and wheat during the waxing moon phases. On the other hand, harvest periods seem to remain independent of the phases of the Earth's natural satellite, but are rather linked to solar cycles. The sequence of agrarian tasks is therefore linked to both the sequence of lunar cycles and the sequence of solar cycles.

In the Cuzco region, the tropical year is divided into two parts: the rainy season and the dry season. The dry season extends from June to the end of November and the rainy season from December to the end of May. The first months of the dry season (June to August) are rather cool, while the following months are hot. Similarly, the first months of the rainy season are hot and the following three months rather cool.

10.4 The Role of the *Sukankas*

Details concerning the organization of the year according to the Inca calendar are given by Bernabé Cobo in his *Historia del Nuevo Mundo* [1653] (Cobo 1956). The solar year began for them on the summer solstice, which falls on 21 December. Years and days were determined from observations of the Sun, while months were quite logically deduced from lunar observations.

According to Cobo, it was the *Sapa Inca Pachakuteq* who organised the transition from the old lunar calendar to another based on the motion of the Sun. The genesis of this new calendar would have begun with the precise measurement of the length of the solar year, using the astronomical pillars built on the heights of Cuzco. All these pillars, called *sukankas*, were used to identify important dates in the calendar by observing the points of sunset or sunrise on certain ridges near the city. These observatories, located in the west to observe the sunsets and in the east to observe the sunrise, were therefore part of 'horizon astronomy,' according to the terminology in use. Cobo mentions that such pillars were also used to indicate where the Sun rose each month. The sighting point for observing these pillars was the Temple of the Sun in Cuzco. According to this same author, pillars were built on the hill called Cinchincalla (the third *waka* on the thirteenth *seq'e* of *Kuntisuyu*), and when the Sun reached them, it was time to start sowing and planting.

Diez de Betanzos [1551] (1987) and Garcilaso de la Vega [1609] (1945, 1969, 2000) provide detailed information on the shape and dimensions of these solar pillars as well as the spacing between them, and there is more about this from Cieza de León [1553] (1984) and by an Anonymous Chronicler [ca. 1570] (1906).

According to Williams (2001), the *sukankas* enabled sufficiently detailed solar observations to determine a precise value for the length of the tropical year. As a result, they could have been used to introduce leap years in the calendar system. The author's reasoning is based on detailed observation of the apparent motion of the Sun. At the time of the equinoxes, around 21 September and 21 March, the points of sunset or sunrise move quickly on the horizon. It has been calculated that the displacement reaches approximately 9.30 m per day on the crests of Picchu where the *sukanka* was located, as viewed from Sabacurinca. From these dates, the daily shift in position of the sunrise and sunset takes place more slowly, until it stops almost completely during the December and June solstices, the position of the Sun remaining almost unchanged for about two weeks.

At the time of the equinox, the Sun fits perfectly between the largest columns of the *sukankas*, but we find that the same phenomenon is not repeated exactly 365 days later. As the length of the tropical year is approximately 365.25 days, the point of sunset moves north in September or south in March. After one year, the displacement is approximately 2.30 m, a quarter of the equinoctial displacement calculated at 9.30 m per day. After another year, or 730 days later, the distance doubles to 4.60 m, and after a third year, or 1095 days, it reaches 6.90 m. But astronomers would undoubtedly have noticed that, 366 days after the last count, the Sun returned to a position such that it could once again fit perfectly between the columns.

The Sun's passages through the Cuzco meridian (zenith) occur on approximately 13 February and 30 October, dates on which the columns or gnomons do not produce a shadow at midday. On these dates, the Sun also moves rapidly over the crests of Picchu, at a rate of approximately 7.80 m per day under the same observation conditions as above.

10.5 Gnomons and Observation of the Equinoxes

Observing the shadow produced by a gnomon at different times of the year made it possible to determine the time of the equinoxes (vernal equinox and autumnal equinox). These observations were carried out by a limited number of people using the *intiwatana* of Machu Picchu or the P'isaq Sun Temple. On the other hand, the passages of the Sun near the markers on the Cuzco horizon which were used to determine the solstices could give rise to spectacular ceremonial rites, because they could be viewed by a larger number of spectators.

In connection with the transition from the Julian to the Gregorian calendar in Peru in 1584, the times of the equinoxes in the original texts of some Spanish chroniclers, including Polo de Ondegardo [1585] (1916), are given as occurring around 11 or 12 March and around 11 or 12 September, dates determined by counting the lunations and observing the positions of the Sun on the heights of Cuzco.

Garcilaso de la Vega [1609] (1945, 1969, 2000) mentions the observation of the March equinox, which corresponds to the corn harvest in the Cuzco region, a special moment celebrated with much pomp, and also the observation of the September equinox. According to this author, the precise moment of the equinoxes was determined by observing the shadow cast by stone columns arranged in the center of circular spaces located near the temples. de Montesinos [1630] (1906) writes that the Incas observed the equinoxes and that these, like the solstices, were used to divide the year into four equal parts. The months corresponding to the vernal equinox and the autumnal equinox were, according to him, called Quilla Toca Torca and Camay Cupac Torca. For his part, de la Calancha [1638] (1981) indicates, although with few details, that the Incas determined the equinoxes from the shadow cast by two pillars placed at the *Qorikancha* in Cuzco.

The writings of Spanish chroniclers attest to the fact that the Incas carefully observed the positions of the Sun on the horizon to identify the places where it rose and set (east and west sides) at the time of the solstices. However, the number and positions of these pillars are not precisely known because there are discrepancies between the various sources as regards this matter. The same goes for the height of the pillars, their separation, and the precise place or places from which the observations were made, although it seems established that the central square of Cuzco was one of these.

An ancient source is given by the writings of Diez de Betanzos [1551] (1987). In accordance with the texts of this author, the Incas year had 360 days and included 12 months of 30 days, approximately 5 days less than the tropical year which began on the December solstice. According to Diez de Betanzos, *Pachakuteq Inca Yupanki* had 8 pillars built on the heights of Cuzco. A group of four pillars was used to indicate the place of sunrise and another group of four pillars was used to indicate the place of sunset. Each group contained two taller external pillars and two smaller internal pillars. From the December solstice, the positions of sunrise and sunset on the horizon gradually move towards the north then towards the south. The Sun therefore rises and sets twice at a specific position on the horizon during the year.

Cieza de León [1553] (1984) mentions the existence of 'solar towers' in Karmenqa (Santa Ana district), northwest of Cuzco. He does not give details about the number of towers or their height. He points out that they allowed the establishment of the calendar by observing the sunsets and he also writes that they were used to measure the length of the shadows produced by the daytime star at different times of the year. The existence of pillars in the Karmenqa district is also confirmed by Antonio Vazquez de Espinosa in his work *Compendio y Descripción de las Indias Occidentales* [1628] (Vásquez de Espinosa 1942), but this author was inspired by previous writings, in particular those of Garcilaso de la Vega.

According to Polo de Ondegardo [1585] (1916), the year consisted of twelve months determined by the phases of the Moon. Each month corresponded to a pillar on the heights of Cuzco. The different months of the year, which began at the winter solstice, were characterized by celebrations in accordance with a decree by *Pachakuteq*. Polo also uses the term *saybas* to designate the pillars in a text entitled *Relación de los fundamentos acerca del notable daño que resulta de no guardar á los indios sus fueros* [1571] (Polo de Ondegardo 1872). According to this author, the Inca year had the same number of days as our own year, except in the case of leap years.

Cobo [1653] (1956) explains that, on the hills surrounding Cuzco, two pillars were placed on the east side and two pillars on the west side at the places where the Sun rose and set at the time of the solstices. As the observations were made from the city of Cuzco, when the Sun appeared at the location of the pillars, this was taken as the beginning of the year. Since the city was located at a latitude of approximately 13 degrees south, this was the time when the Sun rose furthest south. When it rose and set at its farthest point north, it was aligned with the pillars farthest on that side. It then returned to the original point, indicated by the first pillars, and the year was over. The year had twelve months (or moons), the same word *killa* being used to designate the month or the Moon.

Cobo agrees with de Betanzos regarding the start of the year at the December solstice, and that the year had 365 days. According to Cobo, there were a total of fourteen pillars located around Cuzco, some used to indicate the beginning of the months. For de Betanzos, there were four pillars to indicate each solstice while, according to Cobo, there were only two. The Incas observed the phases of the Moon, but if we are to believe de Betanzos, they counted twelve months of thirty days, which quickly introduced a distortion in relation to the lunar year, since the synodic periods (or lunations) count only about 29.5 days. Complementary days had to be included in the thirty-day months to keep the 'calendar year' in line with the tropical year. It cannot be excluded either that extra days may have been added and incorporated into an additional month after a few years, using an intercalation procedure.

We have a text entitled *Discurso de la sucesión y gobierno de los Yngas* by an anonymous author [ca. 1570] (1906). It indicates that there were four pillars on the west side of Cuzco. When the Sun reached the first pillar, it was time to prepare for the plantations in the highlands, and when the position of the Sun coincided with the inner pillars, it was time to think about the plantations in Cuzco, an event that took place during the month of August. This anonymous chronicler also reports that the observation point from which to observe the Sun in relation to the pillars was the *usnu*, located in the main square of Cuzco.

Sarmiento de Gamboa [1572] (1942, 2007) mentions the existence, on the east and west sides of Cuzco, of two sets of four pillars which could have been used to carry out astronomical observations, and in particular to observe the shadows produced by the Sun.

Garcilaso de la Vega [1609] (1945, 1969, 2000) refers to the use of pillars for observing the solstices and for keeping the lunar and solar years in phase. He indicates that there were sixteen pillars divided into groups of four (two on the east side and two on the west side). Given the difficulties involved in observing the Sun from Cuzco, he claims that the outer pillars served to make it easier for observers to locate the inner pillars. This author clearly mentions the simultaneous use of a lunar calendar and a solar calendar, devices kept in agreement by the observation of the solstices. He also indicates that the pillars were still standing on the heights of the city when he left Peru to return to Spain in 1560. According to the writings of de Montesinos [1630] (1882, 1906), it seems established that the pillars were still visible much later, around 1630, approximately a century after the Spanish conquest.

Guamán Poma de Ayala [1615] (1936, 1980) describes how the motion of the sunrise along the horizon was observed during the year, starting from the pillar indicating the December solstice, when the Sun appeared at rest on the horizon for a few days, then moved towards the pillar indicating the point of the June solstice. He also mentions that, for each month, there was an identifiable landmark on the horizon, and that the year was divided into twelve months, in agreement with other chroniclers who allude to the use of lunar and solar observations to define the Inca calendar.

de Molina [ca. 1575] (1989), in his *Relación de las fabulas i ritos de los Incas*, fails to mention the existence of pillars for measuring time. He focuses instead on the description of ceremonies linked to the calendar, notably the festival of *Inti Raymi*, which took place at the June solstice.

10.6 *Mayu*, the Celestial River

The main plane of orientation in the sky to which the Incas referred, and which serves as a basis for contemporary Peruvians speaking Quechua, is not the plane of the ecliptic but that of the Milky Way (*Qolqa Qoyllur*). The Incas did not refer to the same constellations as the Western world. In addition to the asterisms formed from the grouping of bright stars, composing essentially zoomorphic figures, which are sometimes referred to as star-to-star constellations, they also observed dark clouds formed by interstellar dust, obscuring parts of the Milky Way. They assimilated these dark-cloud constellations to animal figures such as a llama, a fox, a snake, and others.

The Milky Way is easily observed in the night sky and plays an important role in the lives of Andean peoples. It bears the name *mayu* ('river') and serves as a benchmark for the definition of the different constellations identified in the sky. Assimilating the Milky Way to a river is not completely illogical. It can in fact be compared to a stream of bright stars flowing against the dark background of the night sky. The terrestrial analogy is provided by the Vilcanota River, flowing from southeast to northwest in the Cuzco region. This river is formed by many small tributaries upstream of the river which converge towards it. Further downstream, it separates into many irrigation canals which bring water to the different cultivated areas. The river therefore passes through a part with maximum flow, comprising a central point which separates the confluence zone from the diffluence zone.

The Andean conception of the Milky Way considers it as a plane rotating around the Earth. This plane is inclined at 26–30° relative to the plane containing the axis of the Earth's poles. As the southeast quadrant of the Milky Way rises, the northwest quadrant descends; when the northeast quadrant rises, the southwest quadrant descends. For an observer in the southern hemisphere, the point where the Milky Way comes closest to the south pole (at 26°) is at the center of the 'Coal Sack,' a very dark region of the sky, and near the star α Crucis, the dominant star of the Southern Cross. This point rotates continuously around a virtual south pole which can be considered the center of the celestial river.

An interesting and very detailed ethnoastronomical study has been carried out by Urton (1981a) in the community of Misminay in the Andes not far from Cuzco. In the cosmology of the people of Misminay, the Earth can be seen as an orange floating on the surface of the water, because it is surrounded by oceans. While terrestrial rivers carry water downward (toward the ocean), the celestial river *mayu* draws water from the oceans and recycles it upward. It is therefore integrated into a process of recycling terrestrial water.

During the dry season, when the Milky Way becomes visible in the night sky, it extends from the northeast to the southwest. At the June solstice which appears during this period, the Sun rises in the northeast. On the other hand, during the rainy season, the Milky Way extends across the night sky from the southeast to the northwest. At the time of the December solstice, which occurs during the rainy season, the Sun rises in the southeast. The celestial river therefore seems closely associated with the Sun since, during the dry season, they both rise in the northeast and, when the rainy season occurs, both are born in the northwest.

10.7 Archaeoastronomy and Ethnoastronomy

Where astronomy and cosmology in the Andes are concerned, it is interesting to distinguish between archaeoastronomy, which studies the calendar and lunar and solar observations during the time of the Incas, and ethnoastronomy, which investigates the cosmology and calendar of contemporary peoples speaking Quechua and inhabiting the Andes.

The Incas did not only observe the Moon and the Sun. They were also keenly interested in the stars. Information on this subject can be found in the dictionaries of the sixteenth and seventeenth centuries and also in the writings of the chroniclers Polo de Ondegardo, Garcilaso de la Vega, Guamán Poma de Ayala, and de la Calancha, discussed in Sect. 3.3. Along with these authors, we should include the anonymous manuscript of Huarochiri and the names of Pachakuteq Yanki Salqamaywa, Avendaño, and Arriaga.

The constellations observed by the Incas are listed by Polo de Ondegardo in his treatise *Los errores y supersticiones de los Indios* [1585] (Polo de Ondegardo 1916). Animals mentioned in this text include birds, llamas, felines, and snakes. According to Urton (1981a, 1981b), this list can be extended to other birds, including the condor, the falcon, and the tinamou, which is almost flightless.

The astronomical observations listed by Urton (1981a, 1981b) and corresponding in particular to the community of Misminay agree with the data mentioned in the Spanish chronicles. As many as 48 individual stars, constellations, or dark clouds appearing in or near the Milky Way are observed by the inhabitants of this locality to determine favorable times for planting and harvesting.

It is amusing to mention here the fact that the Spanish chroniclers had some difficulty in recognizing certain constellations observed in the sky by the Incas, as Garcilaso de la Vega points out. This was because they were looking for constellations composed of individual stars and forming designs or geometric shapes, in the manner of Western astronomy. They were not used to spotting dark constellations due to dust clouds (*yana phuyu*), of which there are several in the southern part of the Milky Way, assimilated to zoomorphic representations.

For a better comprehension of the present chapter, it is worth remembering here that, in contemporary astronomy, 88 constellations are recognized by the International Astronomical Union (IAU), covering the entire celestial sphere. Their names and their boundaries were officially adopted by the International Astronomical Union in 1928 and published in 1930. They are listed in Appendix D.

10.8 Uranometria

In Quechua, planets and bright stars are designated by the term *ch'aska*, while the word *qoyllur* is used for less bright stars. Since ancient times, astronomers have strived to establish catalogues of stars that are easily observable with the naked eye.

Rodrigo Zamorano (1542–1623) was a Spanish cosmographer and mathematician who served King Philip II. He taught in particular at the universities of Valladolid and Salamanca. In his writings from 1585, he mentions the 48 constellations identified by Ptolemy in the *Almagest*. This list of constellations did not include the Southern Cross, a familiar group of stars in the southern hemisphere.

Johann Bayer (1572–1625) was a lawyer passionate about astronomy. In 1603, he published a catalog of the sky called *Uranometria*, which covered the entire celestial sphere. Based on the observations by Tycho Brahe, he included Ptolemy's 48 constellations in his catalog, to which he added a dozen from the southern hemisphere that were unknown to Ptolemy. *Uranometria* is the short title of the atlas of constellations produced by this German astronomer which was published in Augsburg (Germany) by Christophorus Mangus under the full title *Uranometria, omnium asterismorum continens schemata, nova methodo delineata, aereis laminis expressa* (Uranometria, containing charts of all the constellations, drawn using a new method and engraved on copper plates).

Uranometria contains 1005 stars, located with an accuracy of the order of one minute of arc. For his map of the southern sky, Bayer used the observations of two Dutch navigators, Pieter Dirkszoon Keyser and Frederick de Houtman. However, this part of the catalog, which includes 135 stars, turns out to be significantly less precise than the part based on Tycho Brahe's observations, with the positions of the stars sometimes presenting errors of almost two degrees.

The new nomenclature proposed by Bayer designates the stars of a constellation using the letters of the Greek alphabet: α for the brightest star, β for the second brightest, and so on, continuing with the letters of the Latin alphabet if necessary, that is, when the constellation has more than 24 identified stars.

In his catalog, Bayer also divided certain constellations into a large constellation with the same name as the original constellation and a more modest one. This is how the constellation Centaurus was divided into Centaurus and the Southern Cross (the 'Throne of Caesar' in Antiquity). The Southern Cross is mentioned in the text, but does not appear in an illustration.

Bayer thus brought the number of constellations to 62: Ptolemy's 48, to which are added the 12 that he introduced for the southern hemisphere, plus Coma Berenices (Berenice's Hair) and the Southern Cross. These constellations cover the entire celestial vault. Subsequently, additional constellations would be added to those listed by Bayer, notably by Nicolas-Louis de Lacaille (1713–1762), Pierre van der Plancke alias Petrus Plancius (1552–1622), and Johannes Hevelius (1611–1687) in the seventeenth and eighteenth centuries, to reach the 88 constellations currently listed by the International Astronomical Union (see the table in Appendix D).

10.9 The Southern Cross

The Southern Cross, or Crux, is a small constellation in the southern hemisphere, surrounded on three sides by Centaurus and to the south by Musca, the fly. As just mentioned, until the Middle Ages, it was considered as a part of Centaurus. The Southern Cross is useful for finding the celestial south pole. In the absence of a star similar to the pole star of the northern hemisphere, two of the stars of this constellation can be used to locate it: by following the line formed by Gacrux (γ Cru) and Acrux (α Cru) over 4.5 times the distance between these two stars, we arrive at a point very close to the south celestial pole (Urton 1980).

The Southern Cross, aligned with the two feet of Centaurus, is now one of the best-known constellations in the southern hemisphere. It is easily recognized by its characteristic shape, namely four stars of similar apparent brightness forming a Latin cross. The Chakana cross may be an Andean representation of the Southern Cross (Fig. 10.2). The constellation actually contains 49 stars with apparent magnitude less than or equal to 6.51. Its four most easily observed stars, which form the cross itself, are α, β, γ, and δ Crucis. Acrux (α Crucis) is the brightest star in the constellation (see Appendix C). Although appearing to the naked eye as a single star, Acrux is actually a complex star system,

Fig. 10.2 The Chakana cross is mentioned in Spanish chronicles, which indicate its presence in different temples. It may be an Andean representation of the Southern Cross. Author's photograph

comprising several stars. Mimosa (β Crucis), magnitude 1.30, is a blue giant and spectroscopic binary star. Gacrux (γ Crucis) is a red giant. δ Crucis and ϵ Crucis or Ginan are the other two brightest stars in this constellation.

Below the Southern Cross is an example of a dark constellation called the Coal Sack Nebula. The Southern Cross contains an open cluster, also known as the Jewel Box cluster, which contains around a hundred stars. It was discovered by Nicolas-Louis de Lacaille in 1752. In the Cuzco region, it is also called the Calvary Cross in reference to the crosses planted on the tops of mountains, which are invoked for the protection of flocks, to ward off bad weather, and to increase the fertility of the soil.

According to Inca beliefs, the Southern Cross was surrounded by constellations that were directly connected to water and mountains. The appearance of the Southern Cross before the rainy season, when sowing began, and its disappearance after the rains, at harvest time, as well as the observation that it reached its highest point in the sky during the solstice of December, were important in relation to fertility. At Machu Picchu, the fact that the constellation appeared in the center of the Celestial River (the Milky Way) and was aligned with the peak of Salcantay at its highest point had a particular religious and economic significance that was not lost to sky observers at the time.

10.10 The Pleiades

The Pleiades (or the M45 cluster) is an open cluster of stars observable from both hemispheres. It was in 1769 that Charles Messier (1730–1817) added this cluster, visible to the naked eye in the autumn sky, to his astronomical catalog. It is located in the constellation Taurus, near the axis formed by the star Sirius in the constellation of Canis Major (the 'Great Dog'), the Belt of Orion in the constellation of Orion, and the star Aldebaran in the constellation of Taurus.

The Pleiades have been known since Antiquity. The oldest written reference to this constellation dates back to the Greek poet Hesiod (eighth century BCE) for whom the heliacal rising at the beginning of November marked the beginning of winter. The name 'Pleiades' comes from Greek mythology: the Pleiades were seven sisters, the daughters of Atlas and Pleione: Asterope, Merope, Electra, Maia, Taygeta, Celaeno, and Alcyone. Today, several thousand stars have been identified in this cluster, a dozen of which are visible to the naked eye if the weather conditions are favorable. The brightest stars in the cluster take their names from the seven sisters and their parents. Their apparent magnitude is between +2.86 and +6.41, so these stars are visible to the naked eye. The main components of the cluster are given in Table 10.1. See also Appendix C, showing the brightest stars in the sky (Fig. 10.3).

Table 10.1 The brightest stars in the Pleiades cluster. According to Anderson and Francis (2012). List of stars in the Pleiades cluster in XHIP (Extended Hipparchus Compilation). 'Variable' indicates that the magnitude of the star varies

Name	Designation	Apparent magnitude
Alcyone	25 Tauri (or η Tauri)	+2.86
Atlas	27 Tauri	+3.62
Electra	17 Tauri	+3.70
Maia	20 Tauri	+3.86
Merope	23 Tauri	+4.17
Taygeta	19 Tauri	+4.29
Pleione	28 Tauri	+5.09 (variable)
Celaeno	16 Tauri	+5.44
Asterope	21 Tauri	+5.64
	22 Tauri	+6.41
	18 Tauri	+5.65

Fig. 10.3 The Pleiades cluster observed from Mount Palomar in the United States. Credit: NASA, ESA, AURA/Caltech, Palomar Observatory, Public Domain, http://hubblesite.org/newscenter/archive/releases/2004/20/image/a/

It seems clear that the Incas attached great importance to the observation of the Pleiades, designated in particular by the term *Qolqa*, but also by *Oncoy*, *Larilla*, *Fur*, *Pugllaiguico*, and others. These stars were important in relation to the corn harvest, and were celebrated during their heliacal rising at the time of Corpus Christi, a festival of variable date, but generally falling in June.

Observations of the heliacal rising of the Pleiades may have been made from the Temple of the Sun in Cuzco. This possibility was studied in detail by Bauer & Dearborn (1995), in particular in relation to the orientation of the southwest wall of the sanctuary. Taking into account the location of their heliacal rising in the sky around the year 1500, this possibility cannot be excluded, but at the same time no definitive conclusion can be drawn.

Several chroniclers indicate that the Indians observed the Pleiades in the region of the sky which corresponds to the constellation Taurus. This is the case of Polo de Ondegardo [1585] (1916) and also of de Avendaño [1648] (2019), who indicates that the Pleiades were particularly celebrated at the time of Corpus Christi, because it was in this period that the corn was likely to freeze and the crop might be destroyed. This idea is confirmed by Pablo Jose de Arriaga [1621] (1968), who indicates in his well-known work *Extirpación de la Idolatria del Pirú* that the Pleiades cluster appears in the sky at the time of the corn harvest.

According to Bernabé Cobo, the Pleiades, who were also referred to by the word 'Mother,' were honored with special attention all year round and were highly respected by all *ayllus*. The Spanish referred to these stars as *Las Siete Cabrillas* (the seven goats).

10.11 The Constellation of Orion

Orion (the Hunter) is a constellation located almost on the celestial equator. It was already recognised in Antiquity because it is mentioned in particular in the *Odyssey* by Homer, in the poem *Phenomena* by Aratus of Soles, and in the *Aeneid* by Virgil. It was also listed among the forty-eight constellations of Ptolemy's Almagest.

In Greek mythology, Orion was a legendary hunter. Sirius was his dog. Sirius is also the brightest star in the constellation Canis Major. Orion's body is easily visible, being defined by the four bright stars of Rigel (β Ori), Saiph (κ Ori), Betelgeuse (α Ori), and Bellatrix (γ Ori). Orion's belt (or baldric) is made up of the three stars Mintaka (δ Ori), Alnilam (ϵ Ori), and Alnitak (ζ Ori). Saiph (κ Ori) is located on Orion's left knee and is similar in size and distance to Rigel but appears less bright (Fig. 10.4).

Orion is one of the rare constellations that is immediately recognizable by its shape. Its seven brightest stars form an easily identifiable pattern, with four very bright stars defining a characteristic rectangle, in the middle of which is the alignment of three other stars that constitutes its remarkable signature. Betelgeuse (α Orionis), at Orion's left shoulder, is a red supergiant. Rigel (β Orionis), at Orion's right knee, is a blue supergiant and among the brightest stars known (see Appendix C). But the most spectacular celestial object in the constellation is the M42 nebula. Visible to the naked eye (it has magnitude 4.0), it can be seen using binoculars that it is not a star.

Orion's Belt (*Chakana*) is among the constellations that were keenly observed by the Incas. This hypothesis is confirmed by de Avendaño [1648] (2019), who mentions that two constellations were particularly important to them, namely the Belt of Orion and the Pleiades. The stars presented by Polo de Ondegardo [1585] (1916) and by Cobo [1653] (1956) as *Chakana* can be identified with *The Three Marys* (*Los Tres Marias*, in Spanish), which we now call the Belt of Orion, an asterism that is easily recognizable to the naked eye and composed of three blue supergiants. Finally, the Huarochiri manuscript (Huarochiri 1991), a text in Quechua probably written between 1598 and

Fig. 10.4 The constellation of Orion in *Prodromus Astronomia, volume III: Firmamentum Sobiescianum, sive Uranographia*, Table QQ (1690), by Johannes Hevelius (1611–1687). Public Domain

1608, also mentions the observation of three stars forming a straight line and assimilated to a condor, a vulture, and a falcon. This group can quite naturally be identified with Orion's Belt.

10.12 From Urkuchillay to Pihca Conqui

The Incas made no linguistic distinction between planets and bright stars. In Quechua, both were designated by *ch'aska*, while less bright stars were called *qoyllur* (Urton, 1981a).

The most frequently cited sources listing the constellations observed by the Incas are Cobo and Polo de Ondegardo. Both authors refer to a night sky populated by domestic or wild animals. The Incas believed in fact that animals, including birds in particular, had their equivalents in the sky, and that these deities were responsible for their fertility and abundance on Earth. Both mention the group of stars called *Qolqa* (also called *Oncoy*), which corresponds to the Pleiades in the constellation of Taurus, and which were honored everywhere and carefully observed. Their apparent brightness was used to predict the abundance of crops, particularly corn.

There was also a constellation called Urkuchillay which was honored by shepherds and which corresponded to what we call Lyra. This constellation is one of the 48 constellations identified by Ptolemy. Its brightest star is Vega. The Incas thought of it as a llama (*Llama*) of many colors, a deity who would take care of their livestock. They also honored two stars located nearby: the bright one was called Qatachillay and represented, along with a less bright star located not far away in the sky, a female llama and her offspring, or cria. The bright star may possibly correspond to Deneb in the constellation Cygnus and the other to Altair in the constellation Aquila, the Eagle (Bauer & Dearborn 1995).

The forest dwellers also invoked Choquechinchay, which represented a feline and which ensured their protection against jaguars, bears, and pumas. There was also Ancochincay, which protected other animals whose nature was not specified, and Machacuay, responsible for reptiles and snakes, and which protected them from these animals. Finally, the two chroniclers mention other stars which they call *Chakana*, Topatorca, Mamana, Mirco, and Miquiquiray, but without further details. The stars designated by Polo de Ondegardo and by Cobo as Mamana and Mirco, stars associated with a cross which could be the Southern Cross, are perhaps α and β of the Southern Cross. These are among the brightest stars in the sky. A confirmation of this is given by González Holguín [1608] (1989). The object listed as *Chakana* can be identified with Orion's Belt also using Holguín's dictionary. No information is available regarding the stars Topatorca and Miquiquiray. The text written by Polo de Ondegardo was taken up by other authors, including Miguel Cabello de Balboa, José de Acosta, and Antonio de la Calancha.

Like the Sun, stars rise at specific positions on the horizon. A fixed point on the horizon can be used to locate the rising and setting positions of a specific star. During part of the year, a star can be observed to rise during the night, and during the rest of the year it rises during the day and becomes visible in the sky only after sunset. Most stars are not visible during the day when the Sun has the same right ascension. This period of invisibility follows the last evening during which the star is seen after sunset and is called the heliacal setting. The period of invisibility ends with the first morning when the star becomes visible at dawn and is called the heliacal rising. The period of invisibility of the star depends on its brightness, the proximity of its passage near the Sun, the general brightness of the sky, and the transparency of the atmosphere.

According to Bauer & Dearborn (1995), there is no archaeological or astronomical evidence that certain *wakas* of the *seq'e* system developed in the Cuzco region to observe sunrises and sunsets were aligned and used to observe the rising and setting of stars such as the Pleiades, Betelgeuse, β Centauri, etc., as has been affirmed by certain authors (see Zuidema 1982b, for example). The Pleiades, as we know, played an important role in the lives of the Incas because their observation was directly linked to certain constraints of agrarian life, namely, the planting and harvesting of corn. According to studies carried out at the *Qorikancha* in Cuzco, in particular by Zuidema (1982b), it is probable but not certain that these stars were observed from the Temple of the Sun. This does not conflict with a detailed study carried out on the alignments of the different buildings of the Temple of the Sun.

The planet Venus was also the subject of detailed observation and had several names: Ch'aska Qoyllur, Pacaric Ch'aska, Auquilla, Pacari Cuyllor, Chachaquaras, and others.

A recent list of stars observed by the Incas has been compiled by Pacheco et al. (2011). They are given in Table 10.2, where we distinguish the constellations formed by groups of bright stars in the Western fashion and the dark constellations made up of clouds of dust seen against the background of the Milky Way.

10.13 Dark Constellations or *Yana Phuyu*

In the sacred celestial river of the Incas which they called *mayu*, the density of stars is highest around the constellation Sagittarius. In certain places, concentrations of interstellar dust define dark areas which appear in marked contrast against the background of the sky, which is particularly luminous in these regions, given the large number of bright stars (Fig. 10.5).

The Incas identified shapes formed by these dark areas with animals, the most famous undoubtedly being the Llama and its cria. José de Acosta, a Jesuit priest and naturalist, gives a description in his work entitled *Historia natural y moral de las Indias* [1590] (1954). These dark shapes, which were difficult to identify by Westerners arriving in Peru, are referred to in English as 'dark cloud constellations', or just 'dark constellations'.

Among these, we find Yakana between the constellations Scorpius and Centaurus, a constellation mentioned in the Huarochiri manuscript, a text written between 1598 and 1608. This constellation probably included α Centauri (Rigil Kentaurus) and β Centauri (Hadar or Agena), which were seen as representing the eyes of a llama (Llamacñawin). This image of the llama nursing

Table 10.2 Constellations observed by the Incas. It should be noted that the Inca constellations do not necessarily include the same stars as the Western constellations. Some stellar regions only partially overlap. According to Pacheco et al. (2011)

Constellation	Translation	Western equivalent
Bright star constellations		
Willka Wara	Sacred Star	Sirius
Qolla Wara	Star of Qollas	Canopus
K'ancha Wara (Qatachillay*)	Bright Star	Achernar
Choquechinchay	The Golden Feline	Antares
Chuchu Qoyllur (Chukchu Qoyllur)	Star Moving Forward	Aldebaran
Saramama (Saramanka)	Mother of Corn	M7
Qolqa (Qoto)	The Warehouse	Pleiades
Qolqa		Hyades
Urkuchillay	The Little Silver Llama	Lyre
Choqechinchay (Amaru*)	Sacred Serpent	Scorpius
Hatun Chakana (Llaka Unancha Llakachuqui*)	The Great Cross	Orion
Huch'uy Chakana	The Little Cross	The Southern Cross
Thunawa	Batan Grinder	Pegasus
Qolqa		Tail of the Scorpion
Kukamama (Kukamanka)	Mother Coca	Galactic Center
Yakumama	Giant Snake	Tail of the Big Dipper
Dark constellations		
Yakana (Qatachillay)	Sidereal Llama	
Huch'uy Llama	The Llama's Offspring	
Atoq	The Fox	
Michiq	The Shepherd	
Kuntur	The Condor	
Lluthu	The Partridge*	
Hanp'atu	The Toad	
Mach'aqway	The Serpent†	
Ukhumari	The Bear**	
Taruka o Lluych'u	The Deer**	
Urk'uchillay	The Black Llama	

* Two possible interpretations
** Location uncertain
† Different from Amaru

its young is different, however, from the constellation mentioned by Polo de Ondegardo and by Cobo, and should not be confused with it. According to Urton (1981a, 1981b), it can be identified with a dark zone extending between Scorpius and Centaurus.

The author of the Huarochiri manuscript also mentions a dark constellation called Yutu (or Lluthu) which corresponds to the representation of a bird resembling a partridge, called a Tinamou. This is the Coal Sack Nebula, located not far from the Southern Cross. This text also mentions the existence of three stars appearing in a straight line (Orion's Belt) and assimilated to three birds, namely the condor, the vulture, and the falcon, and a group of stars arranged in a circle, called Pihca Conqui. There are other stars for which no description is given, called Poco Huarac, Villca Huarac, and Cancho Huarac.

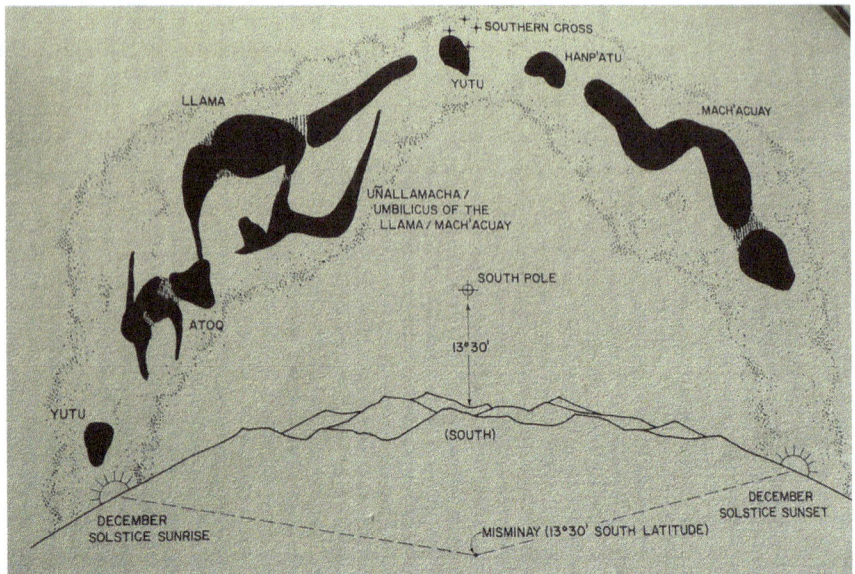

Fig. 10.5 The dark constellations (or *Yana Phuyu*) as visualized in the Misminay community. Illustration taken from G. Urton's *At the Crossroads of the Earth and the Sky* (Urton 1981a), p. 171. Reproduced with the kind permission of the author

If we examine the dark constellations, looking south and moving in order along the Milky Way from west to east, we see the Serpent (*Mach'aqway*), the Toad (*Hanp'atu*), the Tinamou (*Yutu*), the Llama (*Llama*), the Llama's cria (*Uñallamacha*), and the Fox (*Atoq*). From the head of the Serpent to the tail of the Fox, the constellations cover an area of approximately 150° in the sky, lying roughly along the central axis of the Milky Way. It is not generally possible to see all the dark constellations in the sky on the same night. The period of optimal visibility occurs when the region around the Southern Cross and the *Yutu* is on the north–south meridian at midnight, a situation which occurs around 21 March, the autumn equinox in the southern hemisphere. Similarly, the moment of minimum visibility occurs around 21 September, which corresponds to the vernal equinox.

The decrease in contrast of the dark clouds was taken to indicate the approach of rain, while its accentuation would announce a rather dry period. The gradual fading of the dark constellations was in fact an indicator of an increase in atmospheric humidity and a warning of imminent rain. In the minds of the native peoples, the Milky Way, as we have seen, was compared to a celestial river, the dark constellations being located in this river. They therefore constituted a category of celestial phenomena that played an important role in the cosmological relationship between water and the sky, and were taken as important indicators of changes occurring in the physical universe.

As the water in the celestial river was thought to have a terrestrial origin, it is not surprising that the animals appearing in the Milky Way would also be taken to have a terrestrial origin, the dark clouds being parts of the Earth occupied by the Milky Way. It would even be easy to imagine that animals could pass from the Earth into the Milky Way during its transit above the Earth.

In the southern Andes, the rainy season lasts from October to April. Observations show that the Serpent constellation appears in the night sky during this season. Then, during the dry season from May to July, it is located below the horizon and invisible. The same goes for the Llama constellation. The predictions associated with the observation of dark constellations probably took place in August at the start of the sowing period.

Here, we briefly review the different dark constellations and make a few remarks about them.

The Constellation of the Serpent (*Mach'aqway*)

The snake was omnipresent in the Inca world. *Amaru* in Quechua designates a large legendary serpent or dragon from Inca mythology. It had wings and sometimes two heads, a bird's and a puma's. It lived beneath the surface of the Earth, at the bottom of lakes and rivers, and was capable of passing from one world to another. Thus, it could transit from *Hanan Pasha* (the upper or celestial world) to *Kay Pacha* (the terrestrial world), and to *Urin Pasha* (the lower or underground world). This legendary serpent is represented on the Tiwanaku Sun Gate.

Among the Incas, rainbows were likened to giant snakes, the body of the snake forming the arc and the two ends being heads located in springs. The dark constellation Serpent in the Milky Way wraps around Scorpius, this constellation appearing at its zenith in November, at the beginning of the rainy season. The Incas thus imagined a connection between the serpent of the Milky Way, the celestial serpent associated with the rainbow, and the beginning of the rainy season.

The Constellation of the Toad (*Hanp'atu*)

Toads intervene in Andean civilization for divinatory purposes. During the months of September and October, when they are numerous and croaking intensively, it bodes well for abundant rain and excellent harvests. Otherwise, more modest harvests or even a famine should be expected.

The constellation *Hanp'atu* is a cloud of interstellar dust that moves in the Milky Way between the tail of *Mach'aqway* and the Southern Cross. In October, *Hanp'atu* appears in the sky about an hour before sunrise and, thereafter, progressively earlier and earlier than sunrise. This period corresponds to the moment when terrestrial toads come out of their hibernation period and when their croaking is the most intense. The activity of toads is therefore closely linked to agrarian cycles.

The Constellation of Tinamou (*Yutu*)

Near the center of the dark constellations where the Milky Way is close to the south celestial pole, we find, in Quechua terminology, the dark constellation of Tinamou (*Yutu*). This constellation corresponds to the Coal Sack Nebula in Western nomenclature. It is amusing to note that it appears in the sky after the constellation of the Toad, an animal on which this bird feeds.

The constellation of *Yutu* appears in the sky near a group of stars which, in the Western classification, form the Southern Cross. The heliacal rising and setting of α Crucis, the main star of the Southern Cross, occur in early September and around mid-April, respectively, which corresponds roughly to the agricultural season in the Andes. In addition, the constellation *Yutu* passes through the zenith at the time of the December solstice and through the nadir at the time of the June solstice. In this way, the celestial animal is associated with agrarian cycles and solstices.

The tinamou being granivorous, it was appropriate to protect the crops from these birds, as noted by Guamán Poma de Ayala in his writings relating to the Inca period [1615] (1936, 1980). This surveillance lasted from the end of August (sowing) until the beginning of May (harvest), which corresponds to the period when the constellation *Yutu* is visible in the sky.

The Constellation of the Llama (*Llama*)

The llama played a fundamental role in the Andean universe. Used as a beast of burden, it also provided meat for food, wool for making fabrics and clothing, and manure as a soil conditioner. As far as religion was concerned, it was also sometimes offered as a sacrifice. The birth of young llamas took place between December and March, which required special monitoring of the herds during this period to protect them from predators such as foxes.

The dark constellation of the Llama was present in the sky in the period when young llamas were born. On the horizon on the southeast side, we first see the two bright stars α and β Centauri, called in Quechua *llamacñawin* (the eyes of the llama), and this corresponds to the heliacal rising of the constellation at the end of November. Then the rest of the llama's body gradually appears above the horizon, to become completely visible along the north–south meridian at the end of April.

Llamas were connected to the agricultural cycles of the Incas through sacrifices. Polo de Ondegardo mentions in his writings that brown llamas were sacrificed at planting time (from August to September), white and black llamas were sacrificed to obtain rain and abundant herds, and the sacrifice of multi-colored llamas took place at harvest time (from the end of April to the beginning of May).

The Constellation of the Fox (*Atoq*)

The constellation of the Fox is a dark constellation that can be seen perpendicular to the tail of Scorpius, crossing the ecliptic between the constellations Scorpius and Sagittarius. As the Sun moves in the ecliptic plane during its annual revolution, it 'enters' the constellation of the Fox at the time of the December solstice. This is the time when young foxes are born.

The other passage of the Sun 'through' the Milky Way occurs at the time of the June solstice, which corresponds to the breeding period of foxes. It therefore appears that the life cycle of the fox is associated with the passage of the Sun through the solstices and also with the times and places where the Sun's trajectory intersects the Milky Way.

10.14 An Inca Zodiac?

In different civilizations, the notion of a zodiac emerged from observation of the sky. The zodiac is a band approximately 15° wide, located on either side of the ecliptic. This is the region of the sky in which the Moon and the planets of the Solar System move during their annual revolutions. In the zodiac, we find the following constellations: Aries (the Ram), Taurus (the Bull), Gemini (the Twins), Cancer (the Crab), Leo (the Lion), Virgo (the Maiden), Libra (the Scales), Scorpius (the Scorpion), Sagittarius (the Archer), Capricornus (the Goat), Aquarius (the Water-Bearer), and Pisces (the Fish).

The Incas were not familiar with the zodiac as we know it. It has been suggested by Urton (1978) that the Quechua-speaking Andean peoples used a system similar to our zodiac, namely a strip of sky extending from north to south along the Milky Way, this plane of reference being oriented at almost 90° to the plane of the ecliptic. As we know, the Milky Way played an important role in Inca astronomy and beliefs.

The bright star constellations to which the Quechuas refer today, and which were also observed during the time of the Incas, are similar to the ones we recognise insofar as they associate several particularly bright stars considered as forming familiar figures in the sky. Most of these constellations appear close to the Milky Way, and mainly where it reaches its widest near Taurus and Orion. In this region of the sky where the density of stars is lower, bright stars or small clusters characterized by their lower magnitudes are more easily distinguished. This is the case of the Pleiades and the Hyades. The objects associated with these configurations are often architectural elements such as warehouses, bridges, etc. Dark constellations, on the other hand, are formed from concentrations of interstellar dust and appear more contrasted in the regions of the Milky Way where the star density is higher. They are associated with zoomorphic representations.

If we consider the heliacal rising and setting of certain bright stars as they appeared in the sky at the time of the Inca empire, we see that the heliacal risings of Antares, Altair, and Deneb (stars which are found on the left when looking toward the center of the Milky Way) occur around the December solstice, and their heliacal settings around the June solstice. On the other hand, if we consider the stars located on the right when looking toward the Milky Way (this is the case of the Pleiades, Aldebaran, and Sirius), their heliacal risings take place approximately at the time of the June solstice and their settings at the time of the December solstice, in accordance with the fact that, when the first stars rise, the others set.

10.15 *Ch'aska* and the Other Planets

The Incas honored the planet Venus, the Morning and Evening Star, with great fervor, designating it by the word *Ch'aska*. This term is used by Garcilaso de la Vega [1609] (1969, 2000) to refer to this planet, while he uses the term *Qoyllur* to refer to the Pleiades, or to the stars in general. Guamán Poma de Ayala [1615] (1980) uses *Chasca Cuyllor* to refer to a star which is in fact the planet Venus, noting that the Incas did not generally make any linguistic distinction between planets and bright stars.

Several expressions are actually used to refer to Venus. These are listed by Bauer & Dearborn (1995). Thus Guamán Poma de Ayala refers to *Chasca*

Cuyllor or *Pacari Cuyllor*, and Pachakuteq Yanki Salqamaywa to *Chasca Coyllur*. Gonzalez Holguín [1608] (1989) distinguishes the Morning Star (*Chazca coyllur*) and the Evening Star (*Chissichasca*). As for Pablo José de Arriaga [1621] (1968), he attributes two different names to Venus, namely *Pachahuarac* and *Coyahuarac*.

The designations of the other planets are known to us thanks to Valera [ca. 1585] (2009). These are Jupiter (*Pirua*), Mars (*Aucayoc*), Saturn (*Haucha*), and Mercury (*Catu-illa*).

10.16 The Andean Pilgrimage of *Coyllor Riti*

Qoyllur Rit'i (or *Coyllor Riti*, the Snow Festival) is a religious festival held every year in the Andes, in the Sinakara valley south of Cuzco. This is a legacy of local traditions relating to the worship of the stars. In particular, the people of the region celebrate the reappearance of the Pleiades in the sky, associated with the corn harvest and the new year. At this time, the Pleiades have not been visible in the sky since April, but reappear from June onwards. This period also corresponds to the winter solstice (in June), which has been considered since time immemorial as the beginning of the year by many peoples living in the Andes.

During the period when this constellation was absent from the sky, the Incas honored *Pariaqaqa*, the god of water and showers. It then reappeared approximately forty days later, at the time of the harvest, a time of abundance for the people.

The *Coyllor Riti* pilgrimage attracts many representatives of the populations living in the region. It takes place at the end of May or the beginning of June and coincides with the full Moon. These festivities, including processions and dances, take place close to the Christian celebration of Corpus Christi. For indigenous peoples, the culmination of the ceremonies is the sighting of the full Moon (*Killa*), followed by the reappearance of the rising Sun. Until recently, a highlight of this pilgrimage consisted in the 'Ukukus' climbing high up the mountain to bring back a block of ice. They came back down the mountain to Cuzco carrying this block of ice on their shoulders and then participated in the great winter solstice celebration. In Andean mythology, a Ukuku is a creature resulting from the union of a woman with a bear, the latter having transmitted strength and wisdom to her. This tradition has faded since the melting of the glaciers!

10.17 Andean Cosmology and the *Qorikancha*

Juan de Santa Cruz Pachakuteq Yamki Salqamaywa is the author of a drawing which decorates the walls of the *Qorikancha* in Cuzco and which represents a synthesis of Andean cosmology. When viewed from the front, the left side of the drawing is of a masculine inspiration and the right a more feminine one. Among the twenty illustrations present in this composition, six refer to stars or planets (Fig. 10.6).

Fig. 10.6 The drawing by Juan de Santa Cruz Pachakuteq Yamki Salqamaywa which decorates the walls of the *Qorikancha* in Cuzco is a synthesis of Andean cosmology. Author's photograph

In the center and above the central oval, we see a cross formed of five stars: three in a row, with one above and one below. The three horizontal stars are to be compared with the three stars in a straight line mentioned in the Huarochiri manuscript and which can be interpreted as making up Orion's belt (Lehmann-Nitsche 1928). The two additional stars could be Betelgeuse and Rigel, two stars that appear on the list of the ten brightest stars in the sky (see Appendix C). Under the oval, we see another group of five stars: four of them are connected in pairs by lines and the fifth isolated above the cross. Here we read *Chakana* and *Saramama* (the corn goddess). These stars are interpreted as representing the Southern Cross, given the similarity with the pattern formed by the five brightest stars of this constellation.

On the right-hand side of the composition, the female part, we can see the Moon, winter clouds, water, an evening star called Choquechinchay, and a feline also bearing the same name. The position of this animal among the celestial figures suggests that it is the star representing a jaguar that is mentioned in the writings of Polo de Ondegardo and Cobo. On the left-hand side, we can see the Sun, a morning star called Lucero and *Ch'aska Coyllur* (Venus), a group of summer stars, and a star called Qatachillay. The group of stars appearing under the evening star has been interpreted as the Pleiades. Finally, on the left-hand side, we see an illustration of the world, spring, a rainbow, and lightning.

There is only one element which does not allow immediate interpretation. This is Choquechinchay, generally interpreted as a 'crying' feline. This could be a dark constellation, because the observation of such constellations was used to make predictions about the amount of rain, and was thus important for sowing (Urton 1981a). According to this hypothesis formulated by Magli (2005), the problem lies in determining the place in the sky where this constellation might be found. It seems difficult to locate it inside the Scorpion's tail, because this is already occupied by the Fox.

If the Incas identified a dark constellation with the puma, it could be that the layout of Cuzco as a puma is actually a replica of the celestial animal. Cuzco was planned in such a way that the tail of the puma was located on the banks of two rivers, and the same could be true for the celestial puma. In fact, the dark constellations identified by Urton (1981a) are associated with the southern part of the Milky Way, which connects the constellation Scutum (the Shield) to Canis Major (the Great Dog). This part of the galaxy is very bright. In Cuzco, it forms a complete arc in the sky around midnight on the autumnal equinox. The northern part of the Milky Way is, however, clearly visible at the start of the rainy season (from October to November) and appears divided into two branches converging near the constellation Cygnus. Magli (2005) suggests that a possible dark constellation Puma could be located in this part of the Milky Way between the Cygnus (the Swan) and Vulpecula (the Little Fox).

10.18 When Wild Animals Attack the Moon

Eclipses, whether solar or lunar, are among the natural events that have always impressed the human mind. These easily observable events can be dramatic for sky-watchers who do not know what causes them. Among the Incas, as among many other peoples, eclipses generated fear. The total or partial disappearance of the Sun or the darkening of the sky in the middle of the day, or the fact that the lunar disk could be partially or totally darkened, generated anxiety. The fear of the ancients in the face of an eclipse was fueled by the fear that the darkened object would disappear forever.

Total or annular eclipses of the Sun are not very common. Thus Bauer & Dearborn (1995) mention that, between 1440 and 1570, the Inca empire experienced 27 total or annular eclipses of the Sun and approximately twice as many partial eclipses, lunar eclipses being a little less frequent than solar eclipses.

An occultation (or solar eclipse) occurs when the Moon passes in front of the Sun, totally or partially obscuring the latter's disk as viewed from the Earth. To observe this phenomenon, the Moon and the Sun must be in conjunction when observed from the Earth. Total solar eclipses, at a given location on Earth, are fairly rare and short-lived events (less than 8 minutes). When such a phenomenon is visible, the totality is observed only along a narrow band of the Earth's surface which corresponds to the passage of the Moon's shadow. A total eclipse occurs when the Sun is completely obscured by the Moon. During an annular eclipse, the Sun and Moon are perfectly aligned with the Earth, and in this case the Sun appears as a very bright ring surrounding the lunar disk. A partial eclipse occurs when the Sun and Moon are not perfectly aligned and the Moon only partially obscures the Sun.

A lunar eclipse is observed when the Moon is in the Earth's shadow. It takes place when the Moon is illuminated and when the Sun, Earth, and Moon are aligned or almost aligned. A partial lunar eclipse occurs when only part of the Moon enters the Earth's shadow. When the Moon completely passes through the Earth's shadow, the lunar eclipse is total. Unlike a solar eclipse, which may only be seen in a very limited area of the world, a lunar eclipse is visible anywhere on the night side of the Earth. For a lunar eclipse to be observable, the Moon must be near one of the two points of intersection of its orbit with the ecliptic. These are the nodal points, that is, the ascending lunar node and the descending lunar node. Every year, we can generally observe one or two lunar eclipses.

Commentaries on the eclipses observed by the Incas can be found notably in the writings of Garcilaso de la Vega [1609] (1945, 1969, 2000), Cobo [1653] (1956), and Polo de Ondegardo [1585] (1916).

According to Cobo, the appearance of an eclipse was a serious matter. And when it occurred, the indigenous peoples consulted the soothsayers about its meaning. They then made numerous sacrifices of sacred objects, animals, and even human beings. Eclipses could also be interpreted as announcing the death of a prince or an important person. The indigenous peoples claimed that, during a lunar eclipse, a jaguar or a serpent attacked the Moon to devour it. When this phenomenon occurred, it was necessary to make as much noise as possible to scare the feline. They shouted, howled, bashed on pots or drums, made the dogs bark or beat them to make them howl. They said that by making all this noise, they would frighten the jaguar or the serpent and thus prevent it from devouring the relevant celestial object. Otherwise, the world risked being plunged into darkness.

Cobo's text is corroborated by that of Polo de Ondegardo. According to the latter, in fact, when there was a solar or lunar eclipse, people made as much noise as possible to try to ward off the monster which was attacking the day or night star. They also engaged in nocturnal processions around their house with torches lit so that the torments which threatened and which they feared could not occur.

For Garcilaso de la Vega, during a solar eclipse, the inhabitants of the Andes claimed that the star was angry because it had been offended. It seemed to present the face of an angry man and they feared punishment from it. During a lunar eclipse, they claimed that the night star was sick and that if it disappeared, they would all die. The noise they then caused was justified, he said, by the fact that the Moon loved dogs, and if she heard them howling and barking, she would wake from her sleep and heal. Seeing the night star gradually regain its brightness, they imagined that the Moon was healing. When it became bright again, they congratulated each other and were filled with joy.

10.19 Comets and the Sadness of the Inca

Comets are spectacular objects that can be observed anywhere in the sky. They are visible for only a few days or, sometimes, for several months. Several comets can appear in the sky each year but few of them are observable with the naked eye.

Comets were observed in Peru during the time of the Incas. In particular, some of them are mentioned in the texts of Cieza de León [1553] (1984) and Santa Cruz Pachakuteq Yanki Salqamaywa [1613] (1950). Comets are usually unpredictable and sometimes spectacular phenomena. As they were not frequently visible and were not integrated into astronomical cycles, it was quite normal to associate them with exceptional phenomena. They were interpreted by the Incas as omens associated with death.

A comet was visible in the Cajamarca region at the time of the death of *Atawallpa*, who was taken prisoner by Francisco Pizarro on 16 November 1532 and executed on 26 July 1533 (Bernand 2010). In his account, Cieza de León mentions that, when the Inca prisoner saw this comet, he was very affected because he remembered that an omen of the same nature had been observed at the time of the death of his father *Wayna Qhapaq*. This claim is supported by the testimony of Francisco de Xerez, Francisco Pizarro's secretary, who was present in Cajamarca at the time of the emperor's execution. It should be noted that Cieza de León indicates that the death of *Wayna Qhapaq* occurred in 1526.

Santa Cruz Pachakuteq Yanki Salqamaywa describes several comets visible in the Cuzco region at the time of the birth of *Amaru Tupa*, who was the eldest son of the ninth *Sapa Inca*, namely *Pachakuteq Inca Yupanki*. He did not become emperor after the death of his father, because it was his brother who ruled the empire. We cannot identify these comets with certainty, but one of them could have been Halley's comet, which was visible in 1456.

11

Conclusion

Peru is surprising by its extraordinary natural diversity. This manifests itself, among other things, in the scorching expanses of the desert alongside the sometimes turbulent waters of the Pacific when violent phenomena like *El Niño* disrupt its waves; or in the cool waters of Lake Titicaca, near which the Uros peoples have preserved their ancestral traditions; the snow-capped peaks of the Andes, which were once witness to human sacrifices; the heat of the Amazonian jungle overflowing with wildlife; and the cool grassy steppes of the *puna*, where herds of camelids graze. This diversity amazes and cannot leave the traveler indifferent.

A discussion of Andean civilizations, starting with the Inca empire, can only reflect part of the ecological and environmental diversity of this pre-Columbian society, associated above all with the mountainous territories and the harsh climate of the *sierra*. But it is the interactions between the highlands of central Peru and the adjacent lowlands, frequent interactions in the past, which have most contributed to the complexity of this admirable country and its extraordinary artistic and cultural heritage.

The coastal desert stretching along the western edge of the landmass is crossed by numerous rivers that emerge from the foothills of the Andes and flow toward the Pacific. These rivers generate, in the coastal plain, fertile oases where several pre-Columbian civilizations flourished. The interior of the country is essentially made up of the highlands, which once constituted the heartland of the Wari and Tiwanaku cultures. On their eastern slopes, the tributaries of the Amazon, after crossing wild foothills, lose themselves in the lowlands, covered by the luxuriant and almost impenetrable tropical forest.

Interactions favored by the network of waterways have always taken place over the centuries between these three very different ecological zones—the coast, the tropical forest, and the *sierra*—each hosting different ethnic groups, often hostile to one another. It was these three very different regions that the Incas nevertheless managed to dominate, initially through military conquests, before becoming masters of this great unification which led to the formation of their empire, the largest and most astonishing unified territory that has ever existed in South America.

The prestigious and marvelous Inca civilization which developed in these territories around 500 years ago, sometimes cursed by the gods, was the culmination of the cultural contributions of a dozen astonishing and complex pre-Columbian civilizations, such as those, among others, from Chavín de Huántar, Nazca, Wari, or Tiwanaku. The Incas were able to draw on the beliefs and practices of these cultures to establish their own creation and origin myths and thus legitimize the modes of organization of their society, based on hierarchical social, economic, and political relationships.

Inca culture was able to adopt successive contributions from conquered territories and peoples, adapting them to bring about a civilization of extraordinary influence. The integration of civilizational contributions from conquered lands was in fact a key element of the policy of this cultured, agrarian, and warlike people. The Incas adopted the best of the culture from each of the territories they dominated, and their empire appears as a result more like an amalgam of peoples and civilizations than a single monolithic entity.

In the field of art, their ceramics, textiles, and sculpture result from a mixture of local styles with an undeniable Inca flavor. Inca culture was able to assert itself through numerous artistic creations forming a rich and complex iconography of animals, human beings, hybrid creatures, and supernatural representations from previous civilizations. These realistic or anthropomorphic creatures, whose representations are widespread on fabrics and ceramics, have transmitted to us valuable information about local deities and mythical beings predating the Incas, and have also allowed us to improve our knowledge of Inca culture and mythology.

In the field of architecture, among the cultural elements inseparable from the Inca people themselves, we should not neglect the wonderful ingenuity of their craftsmen and builders. The remains of the buildings they erected and which are still omnipresent in the country will surely leave the traveler who discovers them speechless with admiration.

In some cases, the Incas were inspired by ancestral practices inherited from previous civilizations, while in others, they took up the challenge of creating new institutions, developing adaptive strategies, and defining original modes and principles of organization. Ultimately, the genius of their civilization lay in the successful integration of diverse peoples with differentiated traditions and resources into a unified, highly hierarchical, and admirably well organised society. This could only have been achieved by building up from the basic units constituted by the *ayllus*, whose economy was determined by exchanges of goods between members from different ecological zones. Following the processes of conquest and alliance, a high level of bureaucratic organization allowed the state to coordinate and direct the functioning and activities of these different *ayllus*.

The *Sapa Inca* enjoyed considerable prestige because he was considered the earthly representative of *Inti*, the Sun god, whose light and heat contributed to the prosperity of the world. The queen, the *Qoya*, was considered the earthly personification of the goddess of the Moon, a celestial body whose phases punctuated the measurement of time. Around the king and queen, the nobility were distributed in *panakas* dominating the capital Cuzco. Upon leaving this city, officials went to the different districts of the empire to regulate the affairs of state and check that tributes were duly paid in the form of *mit'a* by the *ayllus*. In this way, state control was far-reaching. In the provinces, high-ranking lineages held hereditary lordships and local *kurakas* oversaw state affairs on behalf of the Incas.

A visible mark of the unity established between the power of the capital and the people of the countryside was dramatically expressed by the *Qhapaq Hucha* ceremonies, during which specially chosen victims (usually children) were sacrificed in order to seal the bonds existing between local communities and the Incas of Cuzco, reminding everyone of the unmistakable and well-established hierarchical relationship that existed within the empire. The control of power over the people was also asserted by the *mitmays*, the transfer of populations to different regions of the empire, particularly along its borders.

The history of the Inca people is imbued with an astonishing and omnipresent spirituality. These people in fact attributed metaphysical power to a wide range of objects and a great many places were considered sacred. The objects and places they worshiped constituted one of the foundations of the Inca belief system, and today's peasants in Andean villages continue to revere them. The *wakas* or mystical places were indeed extremely common, particularly in the Cuzco region. Andean religion referred to local beliefs and practices which led to veneration and respect for the spirits of water, the Earth, and the mountains, but also for the deities associated with different ethnic groups and their ancestors scattered across the empire. The ceremonies and ritual practices developed by the Inca nobility contributed to the formation of what we may call an Inca religion, whose goal was the unification of all local groups to ensure Cuzco's hegemony.

The maintenance of mummies was a central and major practice of this religion. The cult of royal mummies and those of the ancestors of the *ayllus* was an important element in the religious life of the empire because, according to widely held beliefs, honoring mummies was essential for the maintenance of cosmic order, and this gesture ensured abundant harvests and promoted the fertility of their animals.

The Andean peoples lived in a world that they considered to be animated by natural but also supernatural forces. The Incas made little distinction between the terrestrial and the celestial, but archaeologists have been able to establish that the locations of certain *wakas* were correlated with the observation of the stars and the Sun in particular. Given the importance it played in regulating the agrarian cycles, the Sun was considered a supernatural power and the *Sapa Inca* was its earthly incarnation.

Observation of the Sun played a key role in the Inca civilization, particularly in determining the privileged moments of the astronomical year in relation to the calendar and the agrarian cycles associated with the tropical year. Solar observations were used to establish the dates of the December and June solstices in relation to the times of sowing and harvesting, and thus heralded major celebrations during the year.

In the field of astronomy, the Incas did not refer to the same constellations as the Western world. In addition to the asterisms formed by groupings of bright stars that compose characteristic figures in the manner of the Western world, they also identified dark constellations in the sky, formed by dust clouds in the Milky Way, which they identified with animal figures. The Milky Way itself, likened to a sacred river called *mayu*, played an important role in the Inca's cosmic conception of the world and served as a major reference point for the definition of the different constellations identified in the sky. Among the stars whose celestial appearances were linked to agrarian cycles, and which were therefore the subject of assiduous observation, we should mention Venus, the 'hairy planet,' the Southern Cross, the constellation of Orion, and in particular the Pleiades.

The traveler who visits present-day Peru will undoubtedly be seduced by magnificent lost cities like the eternal and grandiose city of Cuzco, by its cultural heritage of astonishing richness, by the surprising ruins of distant civilizations, and by the shimmering clothing with sumptuous colors which are the contemporary witnesses to a long tradition in the art of making fabrics and jewelry of surprising beauty, testifying to the ancestral know-how of many generations of artisans.

Peru also has a long, eventful, and fascinating history regarding the many cultures and civilizations that have succeeded one another on this land, so often hostile to humans, but which once attracted many European adventurers and travelers eager to discover the secrets of these peoples, but also eager to appropriate the fabulous riches accumulated by these distant civilizations.

A

Glossary of Spanish (S), Quechua (Q), and Aymara (A)

- **Acsumama** (Q)—Goddess of the potato.
- **Aklla** (Q)—'Chosen Women'. Also called 'Virgins of the Sun', they lived as recluses in convents where they engaged in different activities such as weaving.
- **Aklla Wasi** (Q)—'House of the Chosen Women', a sort of convent housing women in the service of the *Sapa Inca*.
- **Altiplano** (S)—High plateaus of the Andes.
- **Amaru** (Q)—Large snake.
- **Amaru Inca Yupanki** (Q)—Tenth Inca ruler.
- **Amaru Tupaq Inca** (Q)—Eldest son of *Pachakuteq Inca Yupanki* and brother of *Tupaq Inca*.
- **Andenes** (S)—Platforms, terraced crops.
- **Antisuyu** (Q)—Northeastern district of Cuzco and, by extension, the Inca empire.
- **Apu** (Q)—Guardian spirit of a village living on the mountain peaks. Chief.
- **Atawallpa** (Q)—Ruling Inca who was captured and killed by the Spanish at Cajamarca.
- **Atoq** (Q)—Andean fox. Dark constellation.
- **Audiencia** (S)—Courtroom. Architectural feature specific in particular to the Chimú civilization.
- **Ayar** (Ayar Wayqekuna) (Q)—Legendary founders of the Inca dynasty. These four brothers were *Ayar Manqo, Ayar Kachi, Ayar Uchu,* and *Ayar Awqa*. Their respective wives were called *Mama Oqllo, Mama Qora, Mama Rawa* and *Mama Waku*.
- **Ayllu** (Q)—Clan, lineage, community claiming to belong to an ancestor who founded the territory exploited by his descendants.

Appendix A: Glossary of Spanish (S), Quechua (Q), and Aymara (A)

- **Aymara** (A)—Language spoken in Peru and Bolivia (Lake Titicaca region).
- **Ayni** (Q)—Services rendered within the *ayllu*, the community.
- **Cabrillas** (S)—Young goats. The Spanish refer to the Pleiades by the expression *Siete cabrillas*, the Seven Goats.
- **Camellones** (S) or **Waru waru** (Q)—Raised fields separated by canals used for cultivation in the *altiplano* and the *puna*.
- **Campesino** (S)—Peasant, farmer.
- **Caverna** (S)—Type of tomb specific to the Paracas civilization.
- **Ceques** (S)—See *Seq'es* (Q).
- **Ceviche** (S)—Dish made from fish and shellfish.
- **Chakana** (Q)—Crossed lines. Term used to designate Orion's Belt.
- **Ch'aska** (Q)—Shaggy hair. Word used to designate planets, bright stars.
- **Ch'aska Qoyllur** (Q)—Venus or the Morning Star. Goddess of dawn and dusk, protector of young girls.
- **Chaski** (Q)—Person responsible for transmitting messages from relay to relay.
- **Chicha de jora** (S)—Fermented corn-based drink.
- **China** (Q)—Anthropomorphic jars typical of the Chancay culture.
- **Chinchaysuyu** (Q)—Northwestern district of Cuzco and, by extension, the empire.
- **Chullpa** (Q)—Funerary tower specific to certain pre-Columbian civilizations. These are numerous in the Sillustani region.
- **Ciudadela** (S)—Citadel. Fortified city, notably in the Chimú civilization.
- **Corregidor** (S)—Chief Magistrate. Officer of justice.
- **Costa** (S)—Coastal regions.
- **El Niño** (S) ('Little Boy')—Extreme climate episode along the Peruvian coast, characterized by warm water currents ending the fishing season.
- **Encomienda** (S)—Economic system applied by the Spanish throughout the colonial empire during the conquest of the New World.
- **Eqeqo** (Q)—Doll in the form of a peddler. God of abundance and prosperity.
- **Fardo** (S)—A process for treating corpses which consisted in wrapping the body in many layers of fabric to dry it out.
- **Garúa** (S)—Mist appearing on the Pacific coasts.
- **Hanan Pasha** (Q)—Heaven, the upper world in the Inca cosmogony.
- **Hanan Qosqo** (Q)—Upper part of the city of Cuzco.
- **Hanansuya** (Q)—Upper part of the city of Cuzco.
- **Hanp'atu** (Q)—Toad. Dark constellation.
- **Hatun Runa** (Q)—Common people, citizens, mainly peasants.
- **Hawkaypata** (Q)—'Place of Crying' or 'Place of Rest.' A central square in Cuzco.

- **Huaca** (S)—See *Waka* (Q).
- **Huaquero** (S)—Grave robber.
- **Illapa** (Q)—God of lightning and thunder, called Lliwyaq in central Peru.
- **Inca (or Inka)** (Q)—Ethnic group occupying the Cuzco valley, claiming their ancestors to be *Manqo Qhapaq* and *Mama Oqllo*.
- **Inca Roq'a** (Q)—Sixth Inca ruler of the Kingdom of Cuzco. He was the first Inca ruler to bear the title of *Sapa Inca*.
- **Inti** (Q)—Sun god particularly honored by the Incas. He could take three forms: *Apu Inti* (father), *Churi Inti* (son), and *Inti Awqi* (brother).
- **Inti Raymi** (Q)—Ritual celebrations associated with the June solstice.
- **Intiwatana** (Q)—'Anchoring place of the Sun'. Ritual stone table whose legs were oriented towards the cardinal points and which was used in particular in relation to the measurement of time.
- **Kallanka** (Q)—Typically Inca architectural structure consisting of a large rectangular building supported by pilasters. Building used for large gatherings during the Inca era.
- **Kamaq** (Q)—Superior deity believed to project part of its being onto humans.
- **Kancha** (Q)—Typical Cuzco construction comprising a courtyard surrounded by a rectangular enclosure with a single entrance.
- **Kayao** (Q)—Meaning 'original,' characterizing a temple located on a *seq'e*.
- **Kay Pasha** (Q)—World in which we live according to Inca cosmogony.
- **Khipu** (Q)—Device made from knotted ropes intended in particular for storing information. It was used for accounting, but also for recording narratives.
- **Khipukamayoq** (Q)—*Khipu* specialist or master.
- **Killa** (Q)—Moon. Earth's natural satellite. Month.
- **Kinwamama** (Q)—Goddess of quinoa.
- **Kon** (Q)—God of wind and rain.
- **Kukamama** (Q)—Goddess of coca.
- **Kuntisuyu** (Q)—South-eastern district of Cuzco, and by extension of the Inca empire.
- **Kuntur** (Q)—Condor.
- **Kuraka** (Q)—*Cacique*, responsible for the *ayllu*.
- **Kusipata** (Q)—'Rejoicing Square', a central square in Cuzco.
- **K'uychi** (Q)—God of the rainbow.
- **La Niña** (S) ('Little girl')—Climate episode along the coasts of Peru, characterized by cold water currents.
- **Llama** (Q)—Dark constellation of the Llama.

- **Llaqta** (Q)—Community structure larger than the *ayllu*, placed under the authority of a principal cacique (*Apu kuraka*). Sometimes called a province.
- **Lloq'e Yupanki Inca** (Q)—Third Inca ruler of the Kingdom of Cuzco. His existence is semi-legendary.
- **Los Tres Marias** (S)—Belt (or Baldric) of Orion.
- **Lucero** (S)—Bright star.
- **Mach'aqway** (Q)—Dark constellation of the Serpent.
- **Mallki** (Q)—Planted plant. By metaphor, a founding ancestor of an *ayllu* and its territory. By extension, the mummy of this ancestor.
- **Mama Allpa** (Q)—Goddess of fertility.
- **Mama Killa** (Q)—Mother Moon, goddess worshiped particularly on Lake Titicaca. Goddess of marriage and protector of women.
- **Mamakuna** (Q)—Woman in charge of the education of 'chosen girls' in convents.
- **Mama Oqllo** (Q)—Wife of *Manqo Qhapaq*.
- **Mama Qocha** (Q)—Goddess of the sea and fish.
- **Manqo Inca** (Q)—A son of *Wayna Qhapaq* and the half-brother of *Waskar* and *Atawallpa*.
- **Manqo Qhapaq** (Q)—The first mythical Inca.
- **Manto** (S)—Coat. Fabric surrounding a mummy.
- **Mara** (A)—Tropical year.
- **Mayta Qhapaq Inca** (Q)—Fourth *Sapa Inca* of the Kingdom of Cuzco and member of the *Hurin* dynasty.
- **Mayu** (Q)—River. Heavenly river. Term used to designate the Milky Way.
- **Mink'a** (Q)—Collective work carried out for the *ayllu*.
- **Mit'a** (Q)—Work carried out by a household of the *ayllu* for the *cacique* and planned by the imperial authority.
- **Mitma** (or mitimae) (Q)—Foreigner. Colonist moved by the Incas to another province for political or economic reasons.
- **Moiety** (S)—One of two parts of an Andean community, often divided into an 'upper part' and a 'lower part'.
- **Mullu** (Q)—Spondyl: red shell, often having a sacred character.
- **Muyuq Marka** (Q)—Central circular tower of the *Saqsaywaman* fortress.
- **Napa** (Q)—Miniature llama mentioned in the mythical history of the Incas.
- **Necropolis** (S)—Type of tomb specific to the Paracas civilization.
- **Pacha** (Q)—World, Earth.
- **Pachakamaq** (Q)—Creator god, primordial energy at the origin of the cosmos.
- **Pachakuteq Inca** (Q)—Ninth ruler of Cuzco. Often considered to have been behind the expansion of the empire.

Appendix A: Glossary of Spanish (S), Quechua (Q), and Aymara (A)

- **Pachamama** (Q)—Earth goddess in Andean cosmology.
- **Pacsi** (A)—Month or Moon.
- **Panaka** (Q)—Lineage, clan, family group in Inca society.
- **Paqarina** (Q)—Mythical place where the ancestors emerged from the Earth. Place of origin of a lineage, which can be a spring, a cave, or a mountain.
- **Pariaqaqa** (Q)—God of water, showers, and nature in the wild.
- **Payan** (Q)—Secondary: term characterizing a temple located on a *seq'e*.
- **Puna** (Q)—Cold Peruvian highlands with an altitude between 3500 and 5000 m.
- **P'unchaw** (Q)—Gold statue in human form representing the Sun god *Inti*, found in the *Qorikancha* of Cuzco.
- **Qero** (Q)—Kind of wooden or metal vase in which liquids such as chicha were kept.
- **Qhapaq Hucha** (Q)—'Great gift.' Rite practised in exceptional circumstances (critical situations for the Inca) and accompanied by human sacrifices. The entire population was supposed to participate.
- **Qhapaq Inca** (Q)—Noble by heredity.
- **Qhapaq Raymi** (Q)—Festival held in honor of the emperor at the beginning of the year, around the time of the December solstice.
- **Qhapaq Yupanki Inca** (Q)—Fifth *Sapa Inca* of the Kingdom of Cuzco. His predecessor was *Mayta Qhapaq*. He was the last ruler of the Hurin dynasty.
- **Quechua** (S)—Language of the Incas, still spoken in parts of Peru, Bolivia, Ecuador, Chile, and Argentina.
- **Qollana** (Q)—Main: term characterizing a temple located on a *seq'e*.
- **Qollasuyu** (Q)—Southeastern district of Cuzco, and by extension the Inca empire.
- **Qolqa** (Q)—Warehouse, silo, sometimes a circular structure used to store agricultural products. Designation of the Pleiades.
- **Qolqa Qoyllur** (Q)—Milky Way.
- **Qorikancha** (Q)—Sacred enclosure in the center of Cuzco, also called the Temple of the Sun.
- **Qoya** (Q)—Queen. Sister/wife of the *Sapa Inca*.
- **Qoya Raymi** (Q)—Celebration in honor of *Mama Killa*.
- **Qoyllur** (Q)—Star.
- **Qoyllur Rit'i** (or **Coyllor Riti**) (Q)—Andean festival, also called the Snow Festival.
- **Quipu** (S)—See **Khipu** (Q)
- **Raymi** (Q)—Festival.
- **Sapa Inca** (Q)—Supreme power in the Inca empire.
- **Saqsaywaman** (Q)—Fortress, temple located on the heights of Cuzco.

- **Saramama** (Q)—Goddess of corn.
- **Sayri Thupaq Inca** (Q)—Son of *Manqo Inca*. He succeeded the latter after his assassination.
- **Selva** (S)—Amazonian forest.
- **Seq'es** (Q)—System of abstract lines radiating out from the center of Cuzco. On these lines are located *wakas*, or sacred places. In the Cuzco region, 328 *wakas* have been identified, distributed along 41 *seq'es*.
- **Sierra** (S)—Mountain.
- **Sinchi Roq'a Inca** (Q)—Second Inca emperor. Eldest son of *Manqo Qhapaq*, the mythical founder of the Inca empire, and *Mama Oqllo* his sister/wife.
- **Situwa Raymi** (Q)—Festivities organized in Cuzco, at the spring equinox in September, during which a ritual washing of the city took place.
- **Sukanka** (Q)—Solar pillars on the heights of Cuzco, used to determine the dates of the equinoxes and solstices from the observation of sunrises and sunsets.
- **Supay** (Q)—Soul coming out of the body, particularly that of the deceased. Identified in colonial times with the demon or god of death.
- **Suyu** (Q)—District, region of the Inca empire.
- **Tambo** (S)—State warehouse.
- **Tawantinsuyu** (Q)—Name used to designate the Inca empire, the empire of the four regions. It was divided into four parts: *Kuntisuyu*, *Chinchaysuyu*, *Qollasuyu*, and *Antisuyu*.
- **Templo del Sol** (S)—Temple of the Sun (called *Qorikancha* in Cuzco).
- **Titu Kusi Yupanki Inca** (Q)—He was the penultimate *Sapa Inca* of the *Manqo Inca* dynasty. He ascended the throne after the death of his brother *Sayri Thupaq*.
- **Torreón** (S)—Temple of the Sun in Machu Picchu.
- **Tumi** (Q)—Sacrificial knife, often richly decorated.
- **Tupaq Amaru Inca** (Q)—Last legitimate heir to the Inca throne. He was executed in 1572 by Viceroy Toledo.
- **Tupaq Inca Yupanki** (Q)—Emperor of the Incas. Son of *Pachakuteq*.
- **Ukhu Pasha** (or **Urin Pasha**) (Q)—Underworld, or world below.
- **Uñallamacha** (Q)—Dark constellation of the Llama's offspring.
- **Urin Pasha** (Q)—Underworld, lower world, according to Inca cosmogony.
- **Urin Qosqo** (Q)—Lower part of the city of Cuzco.
- **Urinsuya** (Q)—Lower part of the city of Cuzco.
- **Urpu** (Q)—Globular pot with an elongated neck, two handles, and a bulky body, specific to the Wari and Tiwanaku cultures.
- **Usnu** (Q)—Raised place, central platform used during Inca rituals.

Appendix A: Glossary of Spanish (S), Quechua (Q), and Aymara (A)

- **Waka** (Q) (or **Huaca**) (S)—Sanctuary, sacred place, temple where worship was performed. In the Cuzco region, there were several hundred of them distributed throughout the *seq'e* system.
- **Waskar Inca** (Q)—Half-brother of *Atawallpa*, he was the fourth emperor of the Inca empire, succeeding his father *Wayna Qhapaq*, who died of smallpox.
- **Wata** (Q)—Tropical year of approximately 365 days and 6 h.
- **Wayna Qhapaq** (Q)—Last undisputed Inca emperor. Son and successor of *Tupaq Yupanki*. His disappearance caused the civil war between *Atawallpa* and *Waskar*.
- **Willaq Umu** (Q)—High priest of the Temple of the Sun in Cuzco.
- **Wiraqocha** (Q)—*Con Ticci Wiraqocha* was the highest god, the master of the elements. He had two sons, Imaymana and Tocapo. *Wiraqocha* was also the eighth Inca ruler.
- **Yana** (Q)—Servant. Prisoner of war assigned to agricultural tasks.
- **Yana phuyu** (Q)—Dark clouds appearing in the Milky Way.
- **Yawar Waqaq Inca** (Q)—Seventh Inca ruler and the second of the *Hanan* dynasty.
- **Yutu** (Q)—Dark constellation of Tinamou.

B

Glossary of Astronomical Terms

- **Annular eclipse**—An eclipse during which the apparent diameter of the Moon is smaller than that of the Sun. The latter then remains visible in the form of a thin, shiny ring.
- **Apogee**—Point where the Moon, during its trajectory, is furthest from the Earth.
- **Azimuth**—Angle between the meridian and a celestial body, measured clockwise along the horizon from the north.
- **Celestial equator**—Great circle on the celestial sphere, obtained by the intersection of the plane of the terrestrial equator with the celestial sphere.
- **Circadian rhythm**—A biological cycle lasting approximately 24 h. Etymologically, this word comes from the Latin words 'circa', meaning 'around', and 'dies', meaning 'day'. It is therefore a cycle which lasts approximately one day.
- **Comet**—A comet is a celestial body made up of a core of ice and dust orbiting the Sun. The diameter of the core varies between a few hundred meters and a few tens of kilometers. If during its orbit, generally elliptical, the comet transits close to the Sun, the nucleus is then surrounded by a bright atmosphere of gas and dust, called the coma. This extends into a luminous trail of gas and dust which constitutes the tail and which can extend over several million kilometers.
- **Constellation**—Region of the sky containing a certain number of bright stars—bright enough to see with the naked eye—which form a specific pattern in the sky depending on the observer's interpretation. In 1922, the

International Astronomical Union (IAU) divided the sky into 88 constellations, the boundaries of which were precisely defined in 1930.

- **Declination**—Angular distance of a celestial body north or south of the celestial equator. In other words, it is the angle between the celestial equator and a star measured perpendicular to this equator. The declination, which varies between 0 and 90°, takes positive values to the north of the celestial equator and negative values to the south.
- **Ecliptic**—The Earth moves around the Sun and completes an elliptical orbit in one year. In the astronomy of appearances, the Sun moves around the Earth in a plane intersecting the celestial sphere, which is called the plane of the ecliptic. The ecliptic is actually the intersection of the Earth's orbital plane with the celestial sphere, and as the Earth's equator is inclined at an angle of approximately 23° to its orbital plane, the ecliptic plane presents the same inclination relative to the celestial equatorial plane. This inclination defines the obliquity of the ecliptic. The two points of intersection between the ecliptic and the celestial equator correspond to the vernal equinox and the autumnal equinox.
- **Equinox**—The intersection of the plane of the ecliptic with the plane of the celestial equator defines the line of the equinoxes. An equinox or equinoctial point is one of the points of intersection of this line with the celestial equator. We experience two equinoxes during the year, around 21 March and 21 September. The vernal (or spring) equinox is the one occurring in March in the northern hemisphere and the one occurring in September in the southern hemisphere. On the other hand, the autumnal equinox occurs in September in the northern hemisphere and March in the southern hemisphere. At the equinoxes, day and night have equal lengths.
- **Gregorian calendar**—Following the problems with the Julian calendar in use until then, Pope Gregory XIII carried out a reform of the calendar in 1582. This was decided in February of that year with the promulgation of the papal bull *Inter Gravissimas*, and it was implemented in the Catholic States in the month of October: from Thursday 4 October people went straight to Friday 15 October 1582. The leap year system was also reformed.
- **Heliacal rising**—The heliacal rising of a star corresponds to the first day of the year when that star becomes visible in the east, above the horizon at sunrise. Previously, it was hidden below the horizon or located just above it, but invisible due to the luminosity of the Sun. Each day after its heliacal rising, the star appears a little earlier in the sky and remains visible for longer. So-called circumpolar stars do not have a heliacal rising, but are permanently visible above the horizon.

- **Heliacal setting**—The heliacal setting of a star is the last day of the year when that star can be seen above the horizon after sunset. In the days that follow, the star is no longer visible on the horizon. So-called circumpolar stars do not have a heliacal setting, but are permanently visible above the horizon.
- **Hour circle**—The hour circle of a star is the great circle through the star and the two celestial poles.
- **Intercalary month**—A lunar calendar of 12 synodic months or 12 lunations leads to a duration of approximately 354 days. However, the solar or tropical year has a duration of approximately 365 days. To keep the lunar calendar consistent with the solar calendar, an additional month must be added approximately every three years. This is called an intercalary month.
- **Julian calendar**—The Julian calendar was a solar calendar used in ancient Rome. It dates back to 46 BCE and was implemented by Julius Caesar to replace the Republican calendar. It was itself replaced in 1582 by the Gregorian calendar.
- **Lunar eclipse**—When the Earth passes between the Sun and the Moon, the Earth's shadow can be seen projected onto the lunar surface. We then speak of a lunar eclipse.
- **Lunar phases**—The complete cycle of the Moon, during its trajectory around the Earth, called the synodic month or lunation, lasts approximately 29 days 12 h and 44 min. During this cycle, the Earth's natural satellite goes through different phases: new Moon, first quarter, full Moon, last quarter. During the first part of the cycle, the Moon is said to be waxing; during the second part, it is said to be waning. The new Moon is also called conjunction and the full Moon is called opposition.
- **Magnitude**—The brightness of a star is given on a logarithmic scale. The apparent magnitude of an object is a measure of the intensity of radiation received at the Earth in a particular wavelength range. The absolute magnitude of an object is the magnitude that object would have if it were located at a distance of 10 parsecs from the Sun, where 1 parsec is 3.26 light-years. The brighter an object, the lower its magnitude. It increases by one unit if its brightness decreases by approximately a factor of 2.5. The brightest stars in the sky are listed in the table in Appendix C.
- **Mean solar day**—The mean solar day corresponds to the time interval between two successive passages of the mean Sun at the meridian of the location.
- **Milky Way**—The Milky Way, also called the Galaxy, is a collection of stars, gas, and dust. The Solar System is in one of its spiral arms. The Galaxy contains more than 100 billion stars and has the form of a disk with a

diameter of between 80 000 and 100 000 light-years and a thickness of 2000 light-years. Observed from the Earth, the Galaxy appears as a whitish band due to the multitude of stars that cannot be distinguished with the naked eye.
- **Nadir**—Point in the sky vertical to the observer but downward, hence opposite the zenith.
- **Northern hemisphere**—The northern or boreal hemisphere of a planet is the part of it located between the equator and its north pole. In astronomy, this refers to the part of the sky located north of the celestial equator.
- **Obliqueness of the ecliptic**—See ecliptic.
- **Perigee**—Point where the Moon is closest to the Earth during its monthly orbit around the Earth.
- **Precession of the equinoxes**—The Sun and the Moon apply a rotational torque to the Earth, and this results in a change in the orientation of its axis of rotation. The Earth's axis of rotation then describes a small circle in the sky. The precession of the equinoxes is the progressive shift in the direction in which the stars are seen, at a rate of one complete rotation of the Earth's axis of rotation approximately every 25 800 years.
- **Right ascension**—One of the two equatorial coordinates. The other is the declination. The right ascension is the equivalent on the celestial sphere of what we call the longitude on the Earth's surface. It is the angle between the hour circle of the given star and the reference hour circle passing through the vernal equinox. It is measured counterclockwise and expressed in hours, minutes, and seconds, twenty-four hours corresponding to 360°.
- **Sidereal period of the Moon**—The sidereal period corresponds to two passages of the Moon at the same position in the sky in relation to the stars. Its duration is 27.3216609 d or 27 d 7 h 43 min 11.5 s.
- **Sidereal year**—The sidereal year is the time interval which separates two successive passages of the Sun at the same point on the celestial sphere in relation to the fixed stars. Its duration is 365.25636 d or 365 d 6 h 9 min 9 s.
- **Solar eclipse**—A solar eclipse is an occultation produced when the Moon is between the Earth and the Sun, preventing some or all of the sunlight from reaching the terrestrial observer. Eclipses can be partial, total, or annular.
- **Solstice**—The solstices are the points on the ecliptic where the Sun reaches its maximum declination north or south of the celestial equator. The maximum declination towards the north defines the summer solstice and the maximum declination towards the south defines the winter solstice. In the northern hemisphere, the summer solstice occurs around 21 June and the winter solstice around 21 December. In the southern hemisphere, it is the opposite. The solstices correspond to a minimum day length in December

in the northern hemisphere and a maximum day length in June in the same hemisphere.
- **Southern hemisphere**—The southern or austral hemisphere of a planet is the part of it located between the equator and its south pole. In astronomy, this refers to the part of the sky located south of the celestial equator.
- **Stellar cluster**—Our galaxy contains many associations of stars whose members are believed to have a common origin and are linked together by gravitational attraction. Among these, we find in particular globular clusters.
- **Sundial**—The sundial is a simple instrument used for measuring time. It indicates solar time from the shadow projected by an object of variable shape, the style or gnomon, on a surface marked with a series of graduations. This surface is generally flat and horizontal, but it can be vertical, concave, convex, spherical, etc.
- **Synodic period**—The synodic period, also called a lunation, is defined as the time taken for the Moon to return to the same position relative to the Sun. It is equal to 29.5305882 d or 29 d 12 h 44 min 2.8 s.
- **Tropical period**—The tropical period of the Moon is defined as the time between two successive passages of our natural satellite at the same position in relation to the vernal point. It is equal to 27.3215816 d or 27 d 7 h 43 min 4.7 s.
- **Tropical year**—The tropical year is the time interval between two consecutive passages of the Sun observed in the direction of the vernal point. It is equal to 365.24222 d or 365 d 5 h 48 min 48 s.
- **Tropic of Cancer**—The Tropic of Cancer is a parallel located approximately at latitude 23° 26′ north, the northernmost latitude where it is possible to observe the Sun at the zenith during the June solstice. The designation comes from the fact that, in ancient times, the Sun was located in the constellation of the Crab (Cancer in Latin) during the June solstice. Due to the precession of the equinoxes, the Sun is now in the constellation Gemini at the time of the June solstice.
- **Tropic of Capricorn**—The Tropic of Capricorn is the parallel located approximately at latitude 23° 26′ south, the southernmost latitude where it is possible to observe the Sun at the zenith during the December solstice. The designation comes from the fact that, in ancient times, the Sun entered the constellation Capricornus at the time of the December solstice. Due to the precession of the equinoxes, the Sun is now in the constellation Sagittarius at the time of this solstice.
- **True solar day**—The time interval between two consecutive passages of the center of the Sun at the meridian of the location. This varies between 23 h 59 min 39 s and 24 h 00 min 30 s.

- **Zenith**—The zenith is the point of intersection of the vertical of a given location with the celestial sphere. The point on the opposite side is the nadir.

C

Brightest Stars in the Sky

See Table C.1.

Table C.1 List of the brightest stars in the sky (apparent visual magnitude $V < 1.80$). Some stars are variable (var). A first magnitude star is approximately 2.5 times brighter than a second magnitude star. From SIMBAD, the database of the Centre des données stellaires, Strasbourg

	Common name	Designation	Apparent magnitude V
1	Sirius	α CMa	−1.47
2	Canopus	α Car	−0.72
3	Arcturus	α Boo	−0.04
4	Rigil Kentaurus	α1 Cen	−0.01
5	Vega	α Lyr	+0.03
6	Rigel	β Ori	+0.12
7	Procyon	α CMi	+0.38
8	Achernar	α Eri	+0.46
9	Betelgeuse	α Ori	+0.50 (var)
10	Hadar	β Cen	+0.60
11	Capella A	α1 Aur	+0.71
12	Altair	α Aql	+0.77
13	Aldebaran	α Tau	+0.85 (var)
14	Capella B	α2 Aur	+0.96
15	Spica	α Vir	+1.04
16	Antares	α Sco	+1.09

Table C.1 (continued)

#	Name	Designation	Magnitude
17	Pollux	β Gem	+1.15
18	Fomalhaut	α PsA	+1.16
19	Deneb	α Cyg	+1.25
20	Mimosa	β Cru	+1.30
21	Toliman	α2 Cen	+1.33
22	Regulus	α Leo V	+1.35
23	Acrux	α1 Cru	+1.40
24	Adhara	ε CMa	+1.51
25	Shaula	λ Sco	+1.62
26	Gacrux	γ Cru	+1.63
27	Bellatrix	γ Ori	+1.64
28	Elnath	β Tau	+1.68
29	Miaplacidus	β Car	+1.70
30	Alnilam	ε Ori	+1.70
31	Alnitak A	ζ1 Ori	+1.70
32	Alnair	α Gru	+1.74
33	Alioth	ε UMa	+1.76
34	Dubhe A	α1 UMa	+1.79
35	Kaus Australis	ε Sgr	+1.80

D

List of Constellations

See Table D.1.

Table D.1 List of the 88 constellations retained by the International Astronomical Union (IAU). From P. Murdin, *Encyclopedia of Astronomy and Astrophysics*, Vol. I, p. 464, IOP Publishing, Bristol and Philadelphia (2001)

N°	Name	Genitive	Abbreviation
1	Andromeda	Andromedae	And
2	Antlia (The Air Pump)	Antliae	Ant
3	Apus (The Bird of Paradise)	Apodis	Aps
4	Aquarius (The Water-bearer)	Aquarii	Aqr
5	Aquila (The Eagle)	Aquilae	Aql
6	Ara (The Altar)	Arae	Ara
7	Aries (The Ram)	Arietis	Ari
8	Auriga (The Charioteer)	Aurigae	Aur
9	Boötes (The Herdsman)	Boötis	Boo
10	Caelum (The Chisel)	Caeli	Cae
11	Camelopardalis (The Giraffe)	Camelopardalis	Cam
12	Cancer (The Crab)	Cancri	Cnc
13	Canes Venatici (The Hunting Dogs)	Canum Venaticorum	CVn
14	Canis Mayor (The Greater Dog)	Canis Majoris	CMa

Table D.1 (continued)

15 Canis Minor (The Lesser Dog)	Canis Minoris	CMi
16 Capricornus (The Sea Goat)	Capricorni	Cap
17 Carina (The Keel)	Carinae	Car
18 Cassiopea	Cassiopeiae	Cas
19 Centaurus (The Centaur)	Centauri	Cen
20 Cepheus	Cephei	Cep
21 Cetus (The Sea Monster)	Ceti	Cet
22 Chamaeleon (The Chameleon)	Chamaelontis	Cha
23 Circinus (The Compasses)	Circini	Cir
24 Colomba (The Dove)	Columbae	Col
25 Coma Berenices (Berenice's Hair)	Comae Berenices	CrA
26 Corona Australis (The Southern Cross)	Coronae Australis	CrA
27 Corona Borealis (The Northern Cross)	Coronae Borealis	CrB
28 Corvus (The Crow)	Corvi	Crv
29 Crater (The Cup)	Crateris	Crt
30 Crux (The Southern Cross)	Crucis	Cru
31 Cygnus (The Swan)	Cygni	Cyg
32 Delphinus (The Dolphin)	Delphini	Del
33 Dorado (The Goldfish)	Doradus	Dor
34 Draco (The Dragon)	Draconis	Dra
35 Equuleus (The Little Horse)	Equulei	Equ
36 Eridanus (The River)	Eridani	Eri
37 Fornax (The Furnace)	Fornacis	For
38 Gemini (The Twins)	Geminorum	Gem
39 Grus (The Crane)	Gruis	Gru
40 Hercules	Herculis	Her
41 Horologium (The Clock)	Horologii	Hor
42 Hydra (The Water-snake)	Hydrae	Hya
43 Hydrus (The Little Water-snake)	Hydri	Hyi
44 Indus (The Indian)	Indi	Ind
45 Lacerta (The Lizard)	Lacertae	Lac
46 Leo (The Lion)	Leonis	Leo

Table D.1 (continued)

#	Name	Genitive	Abbr.
47	Leo Minor (The Lesser Lion)	Leonis Minoris	LMi
48	Lepus (The Hare)	Leporis	Lep
49	Libra (The Scales)	Librae	Lib
50	Lupus (The Wolf)	Lupi	Lup
51	Lynx (The Lynx)	Lyncis	Lyn
52	Lyra (The Lyre)	Lyrae	Lyr
53	Mensa (The Table)	Mensae	Men
54	Microscopium (The Microscope)	Microscopii	Mic
55	Monoceros (The Unicorn)	Monocerotis	Mon
56	Musca (The Fly)	Muscae	Mus
57	Norma (The Level)	Normae	Nor
58	Octans (The Octant)	Octantis	Oct
59	Ophiucus (The Serpent-bearer)	Ophiuchi	Oph
60	Orion (The Hunter)	Orionis	Ori
61	Pavo (The peacock)	Pavonis	Pav
62	Pegasus (The Winged Horse)	Pegasi	Peg
63	Perseus	Persei	Per
64	Phoenix (The Phoenix)	Phoenicis	Phe
65	Pictor (The Painter's Easel)	Pictoris	Pic
66	Pisces (The Fishes)	Piscium	Psc
67	Piscis Austrinus (The Southern Fish)	Piscis Austrini	PsA
68	Puppis (The Poop or Stern)	Puppis	Pup
69	Pyxis (The Compass)	Pyxidis	Pyx
70	Reticulum (The Net)	Reticuli	Ret
71	Sagitta (The Arrow)	Sagittae	Sge
72	Sagittarius (The Archer)	Sagittarii	Sgr
73	Sculptor (The Sculptor)	Sculptoris	Scl
74	Scorpius (The Scorpion)	Scorpii	Sco
75	Scutum (The Shield)	Scuti	Sct
76	Serpens (The Serpent)	Serpentis	Ser
77	Sextans (The Sextant)	Sextantis	Sex
78	Taurus (The Bull)	Tauri	Tau
79	Telescopium (The Telescope)	Telescopii	Tel
80	Triangulum (The Triangle)	Trianguli	Tri
81	Triangulum Australe (The Southern Triangle)	Trianguli Australis	TrA
82	Tucana (The Toucan)	Tucanae	Tuc
83	Ursa Major (The Great Bear)	Ursae Majoris	UMa
84	Ursa Minor (The Little Bear)	Ursae Minoris	UMi
85	Vela (The Sails)	Velorum	Vel
86	Virgo (The Virgin)	Virginis	Vir
87	Volans (The Flying Fish)	Volantis	Vol
88	Vulpecula (The Fox)	Vulpeculae	Vul

References

Anderson E. & Francis Ch., *Astronomy Letters* **38(5)**, 331 (2012)

Anello Oliva Giovanni, *Historia del reyno y provincias del Perú y varones insignes en santidad de la Compañía de Jesús* [1614], Pontificia Universidad Catolica del Peru, Fondo editorial LXII (1998)

Anonymous Chronicler, *Discurso de la sucesión y gobierno de los Yngas* (ca. [1570]). *Juicio de limites entre el Perú y Bolivia; Prueba peruana presentada al gobierno de la Republica Argentina*, ed. by Victor M. Maúrtua, Vol. **8**, 149–165, Tipografia de los Hijos de M. G. Hernàndez, Madrid (1906)

Antoniadi A. M., *L'Astronomie des Incas et des anciens Péruviens*, L'astronomie **56**, 137–139 (1942)

Ascher M. & Ascher R., *Code of the khipu: A study in media, mathematics and culture*, Ann Arbor, Mich. (1981)

Ascher M. & Ascher R., *Mathematics of the Incas: Code of the Quipu*, Dover Publications (1997)

Astete Victoria F., Ziólkowski M. & Kościuk J., *On the Inca astronomical instruments: the observatory at Inkaraqay-El Mirador (National Archaeological Park of Machu Picchu, Peru)*, Estudios Latinoamericanos **36/37**, 9–25 (2016–2017)

Aveni A. F. (Dir.), *Native American Astronomy*, University of Texas Press, Austin (1977)

Aveni A. F., *Skywatchers of the Ancient Mexico*, University of Texas Press, Austin (1980)

Aveni A. F., *Horizon Astronomy in Incaic Cuzco*. In: *Archaeoastronomy in the Americas*, (ed.), Williamson R. A., A Ballena Press, Center for Archaeoastronomy Cooperative Publication, College Park, Los Altos, pp. 305–318 (1981)

Aveni A. F., *Between the lines. The Mystery of the Giant Ground Drawings of Ancient Nazca, Peru*, University of Texas Press, Austin (2000)
Aveni A. F., *Archaeoastronomy in the Ancient Americas*, J. of Archaelogical Research **11(2)**, 149–191 (2003)
Bandelier A. F. A., *The Islands of Titicaca and Koati*, The Hispanic Society of America, New York (1910)
Bankes G., *Peru before Pizarro*, Phaidon Press, Oxford (1977)
Barbier J.-P., *Guide de l'art précolombien*, Skira (1999)
Baudouin B., *Les adorateurs du Dieu Soleil*, De Vecchi (2002)
Bauer B. S., *The Development of the Inca State*, University of Texas Press, Austin (1992)
Bauer B. S., *The Sacred Landscape of the Inca. The Cusco Ceque System*, University of Texas Press, Austin (1998)
Bauer B. S., *Ancient Cuzco: Heartland of the Inca*, University of Texas Press, Austin (2004)
Bauer B. S. & Dearborn D. S. P., *Astronomy and Empire in the Ancient Andes*, University of Texas Press, Austin (1995)
Bauer B. S. & Stanish C., *Ritual and Pilgrimage in the Ancient Andes*, University of Texas Press, Austin (2001)
Bernand C., *Les Incas: Peuple du Soleil*, Découvertes Gallimard, Paris (2010)
Biémont É., *Rythmes du Temps. Astronomie et Calendriers*, de Boeck, Paris-Bruxelles (2000)
Bingham H., *Lost City of the Incas. With an introduction by Hugh Thomson*, London, Phoenix House (2003)
Bingham H., *Lost City of the Incas. The Story of Machu Picchu and its Builders*, Duell, Sloan & Pierce, New York (1948)
Bingham H., *La Fabuleuse découverte de la cité perdue des Incas: la découverte de Machu Picchu*, transl. from the American by Ph. Babo, with a foreword of D. Lavallée, Pygmalion, 315 p., Paris (1989). New edit., Paris (2008)
Bourget S., *Sacrifice, Violence, and Ideology Among the Moche. The Rise of Social Complexity in Ancient Peru*, University of Texas Press, Austin (2016)
Bray T., *An Archaeological Perspective on the Andean concept of Camaquen: Thinking Through the Late Pre-Columbian Ofrendas and Huacas*, Cambridge Archaeological Journal **19**, 357–366 (2009)
Burger R. L., *Chavín and the Origins of Andean Civilization*, Thames & Hudson, London (1992)
Burger R. L. (ed.), *The Life and Writings of Julio C. Tello: America's First Indigenous Archaeologist*, University of Iowa Press (2009)
Cabello de Balboa Miguel, *Miscelánea antártica, una historia del Perú antiguo* [1586], ed. by L. E. Valcarcel, Instituto de Etnologia, Universidad Nacional mayor de San Marcos, Lima (1951). Online at Kuprienko.info (2010)
Cavatrunci C., Longhena M. & Orefici G., *Pérou des Incas*, Larousse, Paris (2005)
Caviedes C. N., *El Niño in History: Storming through the Ages*, Florida University Press, Gainesville, Fl. (2002)

Changnon S. A. & Bell G. D. (eds), *El Niño, 1997–1998: The Climate Event of the Century*, Oxford University Press, New York/Oxford (2000)

Cieza de León Pedro, *Crónica del Perú: Primera parte* [1553], Academia National de la Historia, Lima (1984)

Cieza de León, Pedro, *Segunda parte de la Crónica del Perú, que trata del señorio de los incas yupanquis y de sus grandes hechos y gobernación*, Imp. by Manuel Gines Hernández, Madrid (1871)

Cieza de León, Pedro, *Tercero libro de las Guerras civiles del Perú el cual se llama la Guerra de Quito*, Imp. by Manuel Gines Hernández, Madrid (1909)

Cobo, Bernabé, *Historia del Nuevo Mundo* [1653]. In: *Obras del P. Bernabé Cobo de la Compañia de Jesús*, ed. by P. Francisco Mateos, Biblioteca de Autores Españoles Vols. **91** et **92**, ed. Atlas, Madrid (1956). Online at Fr.wikipedia.org

Coe M., Snow D. & Benson E., *Atlas de l'Amérique précolombienne*, Édition du Fanal, Amsterdam (1987)

Conrad G. W. & Demarest A. A., *Religion and Empire: The Dynamics of Aztec and Inca Expansionism*, Cambridge University Press, New York (1984)

d'Altroy T. N., *Provincial Power in the Inka Empire*, Smithsonian Institution Press, Washington (1992)

de Acosta, José, *Historia natural y moral de las Indias* [1590]. In: *Obras del P. José de Acosta de la Compañia de Jesús*. Ed. by P. Franciso Mateos, Biblioteca de Autores Españoles Vol. **73**, 3–247, Ediciones Atlas, Madrid (1954) ; *Histoire naturelle et morale des Indes tant occidentales qu'orientales composée en castillan par Joseph Acosta: transl. in French by Robert Regnault Cauxois* [1598]. Online at Gallica

de Albornoz, Cristóbal, *Instrucción para descubrir todas la guacas del Pirú y sus camayos y haziendas* [ca. 1582] In: *Alornoz y el espacio ritual andino prehispanico*, ed. by P. Duviols Revista Andina **2(1)**, 169–222 (1984)

de Arriaga Pablo, José [1621]. *Extirpación de la idolatria del Perú*. In: *Crónicas peruanas de interés indigena*. Ed. by Francisco Esteve Barba, Biblioteca de Autores Espanoles Vol. **209**, pp. 191–277, Ediciones Atlas, Madrid (1968). Online at kuprienko.info (2012)

Dearborn D. & White R., *The 'Torreón' at Machu Picchu as an Observatory*, Archaeoastronomy **5**, Supplement of the Journal for History of Astronomy, S37–S49 (1983)

Dearborn D. & Schreiber K., *Here Comes the Sun: The Cuzco–Machu Picchu Connection*, Archaeoastronomy **IX**, 15–36 (1986)

Dearborn D., Schreiber K. & White R., *Intimachay: A December Solstice Observatory at Machu Picchu*, Peru. American Antiquity **52**, 346–352 (1987)

Dearborn D. & White R., *Inca Observatories: Their Relation to the Calendar and Ritual*. In: Aveni A. (ed.) *World Archaeoastronomy*, Cambridge University Press, Cambridge, pp. 462–469 (1989)

Dearborn D., Seddon M. & Bauer B., *The Sanctuary of Titicaca: Where the Sun Returns to Earth*, Latin American Antiquity **9(3)**, 240–258 (1998)

de Avendaño, Fernando, *Sermones de los Misterios de Nuestra Santa Fe Católica, en Lengua Castellana, y la General del Inca*, Vol. 1: *Impugnanse los Errores Particulares que los Indios han tenido* [1648], Forgotten Books (2019). Online at Archive.org

de Castro Yupangui D., *An Inca Account of the Conquest of Peru*, (Transl. of Ralph Bauer), University of Colorado Press (2005)

Demarest A. A., *Viracocha: The Nature and Antiquity of the Andean High God*, Peabody Museum Monographs, Cambridge, Mass. (1981)

de la Calancha, Antonio, *Corónica moralizada del Orden de San Augustin en el Perú* [1638], ed. by Ignacio Prado Pastor, Universidad National Mayor de San Marcos, Lima (1981). Online at Archive.org

de Matienzo, Juan, *Gobierno del Perú* [1567], Édition et étude préliminaire par G. Lohmann Villena, Travaux de l'Institut français d'études andines XI, Paris-Lima (1967)

de Molina, Cristóbal ('el cuzqueño'), *Relación de las fabulas y ritos de los Incas* (ca. [1575]). In: *Fabulas y ritos de los Incas*, ed. by H. Urbano et P. Duviols, Cronicas de America Series: Historia series **16**, 47–134 (1989). Online at es.wikipedia.org

de Montesinos, Fernando, *Memorias antiguas historiales y politicas del Perú* [1630]. Ed. by Marcos Jimenez de la Espada, Vol. **16**, Imprenta de Miguel Ginesta, Madrid (1882); *Anales del Perú, 1498–1642*, 2 vol., ed. by Victor Manuel Maurtúa y Uribe, Madrid (1906)

de Monzón, Luis (dates unknown), *Descripción de la tierra del repartimiento de los Rucanas Antamarcas de la Corona Real, jurisdicción de la ciudad de Guamanga, ano de 1586* [1586]. In: *Relaciones geográficas de Indias, Perú*, Vol. **1**. Ed. by Marcos Jiménez de la Espada, Ministerio de Fomento, Madrid (1881)

de Murúa, Martin, *Historia general del Pirú* [ca. 1615], Facsimile of J. Paul Getty Museum, MS Ludwig XIII, 16, 804 pp. (2008) *La Historia general del Peru* (1616). Online at biblioteca-antologica.org

de Pasquale N., *The Saved Kingdom* (2011) (http://www.quipus.it/)

de Santa Cruz Pachakuteq Yanki Salqamaywa Juan, *Relación de antiguedades desde Reyno del Pirú* [1613]. In: *Tres relaciones de antiguedades peruanas*, ed. by M. Jimenez de la Espada, pp. 207–281, Editora Guarania, Asuncion del Paraguay (1950)

Diez de Betanzos Juan, *Suma y narracion de los Incas* [1551]. Ed. by Maria del Carmen Martin Rubio, Editiones Atlas, Madrid (1987)

Domenici D., *R.-V. avec l'art précolombien de la Mésoamérique*, Art et civilisation, Rouergue (2009)

du Gourq J., *L'Astronomie des Incas*, La Revue Scientifique, 265–272 (1893)

Dupas A., *Les Civilisations précolombiennes*, Hachette Jeunesse, Paris (2002)

Duviols P., *La lutte contre les religions autochtones dans le Pérou colonial*, Institut d'études andines, Lima (1971)

Espinoza Soriano W., *Los Incas: economia, sociedad y estado en la era del Tahuantinsuyo*, Amaru, Lima (1997)

Evans J., *The History and Practice of Ancient Astronomy*, Oxford University Press, Oxford (1998)

Fabre H., *Les Incas*, 'Que sais-je?', no. **1504**, Presses Universitaires de France, Paris, 8th edn (2005)

García A., *La Découverte et la Conquête du Pérou d'aprés les sources originales*, Paris, Libr. C. Klincksieck (2000)

Garcilaso de la Vega, Inca, *Comentarios reales de los Incas* [1609], 2nd edn, Emecé Editores, Buenos Aires (1945) ; *Royal Commentaries of the Incas and general History of Peru*, Transl. by H. V. Livermore, Univ. of Texas Press, Austin (1969) Ibid., *Commentaires royaux sur le Pérou des Incas*, Paris, Éditions La Découverte (2000)

Ghezzi I. & Ruggles C., *Chankillo: A 2300-Year-Old Solar Observatory in Coastal Peru*, Science **315**, 1239–1243 (2007)

Ghezzi I. & Ruggles C., *The Social and Ritual Context of Horizon Astronomical Observations at Chankillo* In: Ruggles C. (ed.) *Archaeoastronomy and Ethnoastronomy: Building Bridges between Cultures*, Cambridge University Press, Cambridge pp. 144–153 (2011)

Glantz M. H., *Currents of Change: El Niño's Impact on Climate and Society*, Cambridge University Press, Cambridge (1996)

González Holguín, Diego, *Vocabulario de la lengua general de todo el Perú llamada lengua Quichua o del Inca* [1608], Presentación Ramiro Matos Mendieta, Universidad Nacional Mayor de San Marcos, Editorial de la Universidad, Lima (1989). Online at Internet Archive

Guamán Poma de Ayala, Felipe, *El primer nueva Corónica y buen gobierno* [1615]; Travaux et Mémoires de l'Institut d'Ethnologie 23, Université de Paris (1936), ed. by J. V. Murra and R. Adorno and transl. by J. I. Urioste, 3 vol., Mexico City (1980). Online at www5.kb.dk

Gullberg S. R., *The cosmology of Inca huacas*, PhD Thesis, James Cook University (2009)

Gullberg S. R., *Astronomy of the Inca Empire. Use and Significance of the Sun and the Night Sky*, Springer (2020)

Haas J. & Creamer W., *Crucible of Andean Civilization: The Peruvian Coast from 3000 to 1800 B.C.*, Current Anthropology **47**, 745–775 (2006)

Haas J. & Creamer W., *Why Do People Build Monuments? Late Archaic Platform Mounds in the Norte Chico*. In: Burger R. L., Rosenswig R. M. (eds.) *Early New World Monumentality*, University of Florida Press, pp. 289–312 (2012)

Hemming J., *La Conquête des Incas*, Stock, Paris (1971)

Hemming J. & Ranney E., *Monuments of the Incas*, University of New Mexico Press, Albuquerque (1982)

Hernández Astete F., *Las panacas y el poder en el Tahuantinsuyo*, Bulletin de l'Institut français d'études andines **37(1)** (2008)

Hocquenghem A.-M., *L'iconographie mochica et les rites andins: les scènes en relation avec l'océan*, Cahier des Amériques latines **20**, Série sciences de l'homme, 113–129 (1979)

Huarochiri manuscript. See Salomon F., Uriosta G.L., *Huarochiri manuscrit: un testament de la religion andine ancienne et coloniale*, Univ. of Texas Press, Austin (1991). A French transl. by Matthews Taylor is available on a site at Tasmania University: Wikis.utas.edu.au

Hyland S., *Writing with Twisted Cords: The Inscriptive Capacity of Andean Khipus*, Current Anthropology **58(3)**, 412–419 (2017)

Itier C., *Les Incas*, 2nd edn, Les Belles Lettres, Paris (2010)

Karsten R., *La civilisation de l'Empire Inca*, Bibliothèque historique Payot, Paris (1993)
Kauffmann Doig F., *Ancestors of the Incas. The Lost Civilizations of Peru*, (ed.) R. Gheller Doig, Peru (2005)
Kosok P., *Life, Land and Water in Ancient Peru*, Long Island University Press, New York (1965)
Kosok P. & Reiche M., *The Mysterious Markings of Nazca*, Natural History **56**, 200–207 (1947)
Krupp E. C., *In Search of Ancient Astronomies*, McGraw-Hill, New York (1979)
Krupp E. C., *Echoes of the Ancient Skies*, Harper & Row, New York (1983)
Lambers K., *The geoglyphs of Palpa (Peru)*, PhD Thesis, University of Zürich (2004)
Larco Hoyle R., *Los Mochicas*, Tomes I et II, Vol. I, Lima (Peru) (1938); Vol. II, Lima (Peru) (1940)
Lavallée D. & Lumbreras L. G., *Les Andes: de la Préhistoire aux Incas*, Univers des formes, Paris (1985)
Lehmann-Nitsche R., *Coricancha: El Templo del Sol en el Cuzco y las imagenes de su altar mayor*. Revista del Museo de La Plata **31**, 1–256 (1928)
Longhena M., *Les Incas*, Gründ, Paris (1999)
Longhena M. & Alva W., *Les Incas: les civilisations andines, des origines aux Incas*, Gründ, Paris (1999)
Magli J., *Nexus Network Journal* **7(2)**, 22–32 (2005)
Malville J. M., *Cosmology in the Inca Empire: Huaca Sanctuaries, State-Supported Pilgrimage, and Astronomy*, Journal of Cosmology **9**, 2106–2120 (2010)
Malville J. M., *Pre-Inca Astronomy in Peru*, In: *Handbook of Archaeoastronomy and Ethnoastronomy*, Springer Science & Business Media, New York, p. 795 (2015)
Malville J. M. , Thomson H. & Ziegler G., *The Sun Temple of Llactapata and the Ceremonial Neighborhood of Machu Picchu*. In: Malville M. J., *Cosmology in the Inca Empire: Huaca Sanctuaries, State-Supported Pilgrimage, and Astronomy*, Journal of Cosmology **9**, 2106–2120 (2010)
Marín-Dale M., *Decoding Andean Mythology*, Salt Lake City: University of Utah Press (2016)
Mason J. A., *The Ancient Civilizations of Peru*, Penguin, New York (1968)
Mathé A. & Mathé P., *Terre des Incas*, (ed.) A. Barthélemy, Avignon (1996)
Métraux A., *Les Incas*, Seuil, coll. 'Points Histoire', Paris (1983)
Métraux A., *Religions et magies indiennes d'Amérique du sud*, Gallimard, Paris (1967)
Molinié-Fioravanti A., *La Vallée sacré des Andes*, Société d'ethnographie, Paris (1982)
Moseley M., *The Incas and Their Ancestors*, Thames & Hudson, London (1992)
Niles S., *The Shape of Inca History: Narrative and Architecture in an Andean Empire*, University of Iowa Press, Iowa City (1999)
Orlove B. S., Chiang J. C. & Cane M. A., *Forecasting Andean rainfall and crop yield from the influence of El Niño on Pleiades visibility*, Nature **403**, 68–71 (2000)
Pacheco E., Flores S. & Salazar E., *Some notes on the Inka constellations* In: D. Valls-Gabaud & A. Boksenberg (eds.) *The Role of Astronomy in Society and Culture*, Proceedings IAU Symposium No. 260 (2011)

Pareja D., *Pre-Hispanic tools of computation: the quipu and the Yupana*, Rev. Integr. Temas Mat **4(1)**, 37–56 (1986)

Pärssinen M., *Tawantinsuyu. El estado inka y su organización política*, Institut français d'études andines, Pontificia Universidad Católica del Perú, Lima (2003)

Pease F. G. Y., *Histoire des Incas*, Maisonneuve & Larose (1995)

Pizarro, Pedro, Relación del descubrimiento y conquista del Perú [1571]; *Récit de la découverte et de la Conquête des royaumes du Pérou*, Éditions du Félin, Paris (1992)

Polo de Ondegardo, Juan, *Relación de los fundamentos acerca del notable daño que resulta de no guardar a los indios sus fueros* [1571], Colección de documentos inéditos relativos al descubrimiento, conquista y colonización de las posesiones españolas en América y Oceanía, CODOIN-Indias, Vol. 17, 7–100 et 101–177 Madrid (1872). Copy from the Peabody Library at Harvard University, Internet archives

Polo de Ondegardo, Juan [1585], *Los errores y supersticiones de los Indios, sacadas del tratado y averiguación que hizo el Licenciado Polo*. Éd. par Horacio H. Urteaga & Carlos Romero, Colección de Libros y Documentos Referentes a la Historia del Perú, Vol. 3, pp. 3–43, San Marti y Ca, Lima (1916). Online at Es.wiki.org

Pozzi-Escot D., Santa Gadea L. M. & Amico J. C., *Pachacamac, El oraculo en el horizonte marino del sol poniente*, Colecion Arte y Tesoros del Peru (2017)

Prescott W. H., *Aztèques et Incas. Grandeur et décadence de deux empires fabuleux*, Pygmalion, Paris (2007)

Prescott W. H., *History of the Conquest of Peru*, Dover Publications, New York (1847); reprinted (2005)

Protzen J. -P., *Inca Architecture and Construction at Ollantaytambo*, Oxford University Press, Oxford (1993)

Quilter J. & Urton G. (eds.), *Narrative Threads, Accounting and Recounting in Andean Khipu*, University of Texas Press, Austin (2002)

Rachowiecki R. & Beech Ch., *Pérou*, 2nd edn, Lonely Planet, Paris (2006)

Radicati di Primeglio C., *El Sistema Contable de los Incas. Yupana e Quipu*, Libreria Studium, Lima, Peru (1976)

Reiche M., *Mystery on the Desert*, Offzindruck AG, Stuttgart (1968)

Reinhardt J., *Sacred Mountains: An Ethno-Archaeological Study of High Andean Ruins*, Mountain Research Development **5(4)**, 299–317 (1985)

Reinhardt J., *Discovering the Inca Ice Maiden*, National Geographic Society, Washington D.C. (1998)

Reinhardt J., *The Ice Maiden: Inca Mummies, Mountain Gods, and Sacred Sites in the Andes*, National Geographic Society, Washington D.C. (2005)

Reinhardt J., *Machu Picchu, The Sacred Center*, 3rd edn, Nuevas Imágenes S.A., Lima (2007)

Reiss W. & Stübel A., *The Necropolis of Ancon in Peru: a contribution to our knowledge of the culture and industries of the empire of the Incas*, 3 vols, Asher & Co, Berlin (1880–1887)

Rick J., *Context, Construction, and Ritual in the Development of Authority at Chavín de Huantar*. In: Conklin W. J., Quilter J. (eds.) *Chavín: Art, Architecture, and Culture*, Cotsen Institute of Archaeology, Los Angeles, p. 3 (2008)

Rowe J. H., *Inca Culture at the time of Spanish conquest*, in: *Handbook of South American Indians*, Vol. 2, Smithsonian Institution, Washington (1946)

Rowe J. H., *The Kingdom of Chimor*, Acta Americana **6** (nos. 1 et 2), 26–59 (1948)

Rowe J. H., *The Incas under Spanish Colonial Institutions*, Hispanic American Historical Review **37(2)**, 155–199 (1957)

Rowe J. H., *Max Uhle, 1856–1944. A Memoir of the Father of Peruvian Archaelogy*, University of California Publication, Vol. **46** (1954)

Roza G., *Incan Mythology and Other Myths of the Andes*, The Rosen Publishing Group, Inc. (2008)

Salazar L., *Machu Picchu: Unveiling the Mystery of the Incas*, (eds.) R. Burger & L. Salazar, Princeton University Press, New Haven, pp. 21–47 (2004)

Salazar F. E. E. & Salazar E. E., *Cuzco and the Sacred Valley of the Incas*, Tankar E. I. R. L., Cuzco, Peru (2014)

Salomon F. & Urioste G. L., *The Huarochiri Manuscript: A Testament of Ancient and Colonial Andean Religion*, University of Texas Press, Austin (1991)

Sarmiento de Gamboa, Pedro, *Historia de los Incas* [1572], Emecé Editores, Buenos Aires (1942); *The History of the Incas*, transl. and ed. by B. S. Bauer & V. Smith, Univ. of Texas Press, Austin (2007)

Scott L. C., *Gods, Goddesses, and Mythology*, Vol. II, Marshall Cavendish Press (2005)

Shady Solis R., *America's First City? The Case of Late Archaic Caral*, In: Osbell W. H. & Silvermann H. (eds.) *Andean Archaeology III: North and South* Springer, New York, pp. 28–66 (2006)

Staller J. E., *Dimensions of Place: The Significance of Centers to the Development of Andean Civilization: An Exploration of the Ushnu Concept*, In: Staller J. E. (ed.) *Pre-Columbian Landscapes of Creation and Origin*, Springer, New York, pp. 269–313 (2008)

Steele P. R. & Allen C. J., *Handbook of Inca Mythology*, ABC-CLIO, Inc. (2004)

Uhle M., *Die Ruinen von Moche*, Journal de la Société des Américanistes, pp. 95–117 (1913)

Uhle M., *Las Antiguas Civilizaciones del Peru frente a la Arqueologia e Historia del Continente Americano*, Sobretiro de la 'Revista del Museo Nacional', Tome **XXV**, Lima (1956)

Uhle M. & Stübel A., *Die Ruinenstaette von Tiahuanaco im Hochlande des alten Peru*, C.T. Wiskott, Breslau (1892)

Urton G., *Beasts and Geometry: Some Constellations of the Peruvian Quechuas*, Anthropos. **LXXIII**, 32–40 (1978a)

Urton G., *Orientation in Quechua end Incaic Astronomy*, Ethnology **XVII(2)**, 157–167 (1978b)

Urton G., *Celestial Crosses: The Cruciform in Quechua Astronomy*, J. of Latin American Lore **VI(1)**, 87–110 (1980)

Urton G., *At the Crossroads of Earth and Sky: An Andean Cosmology*, University of Texas Press, Austin (1981a)

Urton G., *Animals and Astronomy in the Quechua Universe*, Proceedings of the American Philosophical Society **CXXV** (2), 110–127 (1981b)

Urton G., *From knots to narratives: reconstructing the art of historical record keeping in the Andes from the Spanish transcriptions of Inka Khipus*, Ethnohistory **45**, 409–438 (1998)

Urton G., *Signs of the Inka Khipu*, University of Texas Press, Austin (2003)

Urton G., *Mythes incas*, Éditions du Seuil, Points, Paris (2004)

Urton G., *Signs of the Inka Khipu: Binary Coding in the Andean Knotted String Records*, University of Texas Press, Austin (2005)

Valera, Blas, *Relación de las Costumbres antiguas de los Naturales del Piru* ([ca. 1585]). In: *Tres relaciones de antiguedades peruanas*, ed. by M. Jimenez de la Espada, 135–203, Editora Guarania, Asuncion del Paraguay (1950). Online at Kuprienko.info (2009)

Van de Guchte M. J. D., *Carving the World: Inca Monumental Sculpture and Landscape*, PhD Thesis, University of Illinois (1990)

Vásquez de Espinosa, Antonio, *Compendio y Descripción de las Indias Occidentales* [1628], transl. by Charles Upson Clark, Smithsonian Miscellaneous Collections, Smithsonian Institution, Washington D.C. (1942)

de Velasco y Pérez Petroche, Juan, *Historia moderna del Reino de Quito y crónica de la provincia de la Compañía* (1789), reedited in Quito, Editorial de la Casa de la Cultura Ecuatoriana (3 vols) (1979)

Villacorta Ostolaza L. F., *Ceramics of Ancient Peru*, Gheller Doig R. (ed.), Peru (2007)

von Hagen V., *Le Pérou avant les Incas*, Éd. France-Empire, Paris (1979)

Waisbard S., *Machu Picchu, fabuleuse cité perdue des Incas*, Robert Laffont, Paris (1976)

Williams C., *Sucancas, Quipus y Ceques. El tiempo y la sacralización del espacio en el Cusco*, Revista del Museo National del Peru, Lima, Peru, Tome **XLIX**, 123–162 (2001)

Wright K. & Valencia A., *Machu Picchu: A Civil Engineering Marvel*, ASCE Press, Restone VA (2000)

Zawaski M. J., *Archaeoastronomical Survey of Inca Sites in Peru*. MA Thesis, University of Northern Colorado, Greeley-ProQuest Dissertations and Theses (2007)

Zawaski M. & Malville J. M., *An Archaeoastronomical Survey of Major Inca Sites in Peru*, Archeoastronomy: The Journal of Astronomy in Culture **XXI**, 20–38 (2007–2008)

Ziólkowski M., Kościuk J. & Astete Victoria F. *Astronomical Observations at Intimachay (Machu Picchu): A New Approach to an Old Problem*, Anthropological Notebooks **XIX** (Supplement) (2013)

Ziolkowski M., Kosciuk J. & Victoria F.A. *Astronomical observations at Intimachay (Machu Picchu): A new approach to an old problem*. In: Sprajc I. & Pehani P., *Ancient Cosmologies and Modern Prophets*. Ljubjana, Slovenia: Slovene Anthropological Society pp. 391–404 (2013)

Ziolkowski M. & Kosciuk J., *Astronomical Observations in the Inca Temple of Coricancha (Cusco)? A Critical Review of the Hypothesis*, Polska Akademia Nauk Oddzial W Lublinie, Tome **XIV/1**, Lublin (2018)

Zuidema R. T., *The Ceque System of Cusco: The Social Organization of the Capital of the Inca*, Brill, Leiden (1964)

Zuidema R.T., *The Inca calendar*, in: Native American Astronomy, Aveni A. F. (ed.), University of Texas Press, Austin (1977)

Zuidema R. T., *Lieux sacrés et irrigation: tradition historique, mythes et rituels au Cuzco*, In: Annales, Economies, Sociétés, Civilisations 33ème année **5–6**, 1037–1056 (1978)

Zuidema R. T. *Inca Observations of the Solar and Lunar Passages Through Zenith and Anti-Zenith at Cuzco*, In: Archaeoastronomy in the Americas, Williamson R. A. (ed.), A Ballena Press/Center for Archaeoastronomy Cooperative Publication, College Park, Los Altos. pp. 319–342 (1981)

Zuidema R. T., *The sideral lunar calendar of the Incas*, in: Archaeoastronomy in the New World, Aveni A. F. (ed.), Cambridge University Press, Cambridge London New York, pp. 59–107 (1982a)

Zuidema R. T., *Catachillay. The Role of the Pleiades and the Southern Cross and Centauri in the Calendar of the Incas*. In: Ethnoastronomy and Archaeoastronomy in the American Tropics, Aveni A. F. & Urton G. (eds.), Annals of the New York Academy of Sciences **385**, 203–229, New York (1982b)

Zuidema R. T., *The Place of the Chamay Wariqsa in the rituals of Cusco*, Amérindia **11** (1986)

Zuidema R. T., *El sistema de ceques del Cuzco: La organización social de la capital de los Incas*, Pontificia Universidad Católica del Perú (1995)

Zuidema R. T., *Pilgrimage and Ritual Movements in Cusco and the Inca Empire*. In: Malville J. M. & Saraswati B. (eds.) *Pilgrimage: Sacred Lanscape and Self-Organized Complexity*, New Delhi: Indira Gandhi National Center for the Arts, 269–288 (2008)

Zuidema R. T., *Tiwanaku Iconography and the Calendar* In: Young-Sanchez (ed.) *Tiwanaku* Denver Art Museum, Denver, pp. 82–100 (2009)

Zuidema R. T., *El Calendario Inca. Tiempo y Espacio en la organización Ritual del Cuzco; la Idea del Pasado*, Fondo Editorial del Congreso del Perú, Fondo Editorial Pontificia Universidad Católica del Perú, Lima (2010)

Zuidema R. T., *Ceque System of Cuzco: A Yearly Calendar-Almanac in Space and Time*. In: *Handbook of Archaeoastronomy and Ethnoastronomy*, Ruggles C. L. N. (ed.), Springer Science & Business Media, New York, pp. 851–863 (2014)

Zuidema R. T. & Quispe M. U., *A visit to God – The account and interpretation of a religious experience in the Peruvian community of Choque-Huarcaya*, Bijdragen tot de Taal-Land en Volkenkunde **124**, 22–39 (1973)

Subject Index

A

Adobe house, 41, 57, 59, 62, 109, 126
Agrarian cycles, 36, 159–161, 182, 194
Agricultural terraces, 104
Almagest, 170, 175
Alpaca, 6, 8, 10, 28, 41, 42, 54, 58, 83, 118, 119, 126
Altiplano, 34, 54, 70, 72, 81, 88, 126, 127, 198
Amazon forest, 1, 5, 34, 202
Anatomical implausibility, 89
Andean agriculture, 80
Andean civilizations, 33–35, 143, 181, 191
Andean clan (ayllu), 1, 67, 68, 79, 82, 83, 105, 135, 139, 150, 175, 193, 194, 197–200
Anthropology, 14, 16
Archaeoastronomy, 151, 169
Archaeological sites, 2, 14, 18, 31, 86
Archaeology, 13–18, 37, 39, 43, 44
Archaic period, 19, 145
Art of the textiles, 41, 42, 192

Astronomical observations, 2, 46, 47, 123, 124, 154–156, 159, 167, 169
Audiencias, 28, 60, 197
Autumnal equinox, 130, 164, 165, 187, 206
Aymara language, 53, 65, 133, 161, 197, 198

B

Bath of the Inca, The, 101
Beer (chicha), 56, 61, 82, 99, 150, 162, 198, 201
Bright stars, 2, 156, 168, 170, 175–178, 183, 184, 194, 198, 205

C

Cacique (kuraka), 1, 79, 193, 199, 200
Camelids, 2, 8, 28, 56, 58, 118, 191
Cataclysmic destruction of the world, 67, 84, 85

Subject Index

Celestial river (Mayu), 168, 178, 194, 200
Central Andes, 5, 6, 8, 18, 34, 56
Ceramics, 2, 14, 16, 17, 19, 20, 36–41, 43, 44, 48, 50, 52, 55, 58, 59, 61, 63, 70, 89, 192
Ceremonial center, 2, 19, 34–37, 40, 41, 73, 90, 146
Chacana cross, 103
Chancay culture, 35, 61, 198
Chavín de Huántar culture, 1, 17, 20, 34–38, 90, 91, 99, 146, 192
Chimú culture, 14, 17, 35, 57, 197, 198
Chincha culture, 34, 35, 61
Chosen women (akllas), 197
Chullpa, 20, 198
City of Vilcabamba, 15, 75, 113–115

Ciudadelas, 57, 59, 60, 198
Conquistadors, 2, 20, 22–24, 28, 75, 96, 106, 114
Constellation of Orion, 160, 173, 175, 176, 179, 184, 194, 215
Constellation of the Fox, 124, 183, 168, 179, 180, 197, 215
Constellation of the Llama, 147, 168, 180–183, 199, 202
Constellation of the Serpent, 179–181, 200
Constellation of the Tinamou, 124, 169, 179, 180, 182, 203
Constellation of the Toad, 124, 125, 179, 180, 182, 198
Constellations, 2, 46, 47, 61, 83, 84, 123–125, 130, 148, 149, 156, 168–185, 187, 194, 205, 206, 209, 213
Cordillera de los Andes, 6, 9, 77, 84
Costa, 1, 198
Creative triad, 66
Creator god, 66–68, 86, 108, 115, 118, 200
Crop calendar, 150

Crossed hands temple, 18
Cult of the sun, 87, 94, 143
Cupisnique civilization, 17, 34, 35, 37–39
Cuzco valley, 69, 70, 77, 93, 101, 136, 199
Cyclopean statuary, 53

D

Dark constellations, 147, 170, 172, 178–184, 187, 194, 197, 198
Districts of Cuzco, 71, 77, 96, 135, 136, 197-199, 201
Drawing by Juan de Santa Cruz Pachakuteq Yamki Salqamaywa, 186

E

Early (or Ancient) Horizon, 34, 35, 91, 145
Early (or Ancient) Intermediate period, 15, 34, 35, 40, 56, 70
Ecliptic, 129, 130, 152, 168, 183, 184, 188, 206, 208
El Niño, 3, 9–11, 19, 37, 41, 43, 91, 149, 191, 198
El Señor de Ucupe, 50
El Señor de Sipan, 19, 49, 50
Empire of the four directions, 77
Encomienda, 21, 115, 198
Equinoctial alignments, 102
Equinoxes, 2, 46, 47, 99–101, 123, 130–132, 148, 152–154, 156, 159, 161, 164, 165, 202, 206, 208, 209
Ethnoastronomy, 169
Excavation campaigns, 15
Extirpation of idolatry, 2, 30, 82, 115, 135, 143, 147

F

Fardos of Paracas, 39, 56, 62
Festival of the sun, 98
Five suns, The, 67
Fortress of Saqsaywaman, 73, 78, 93, 95–99, 106, 148, 200, 201
Funeral centers, 39

G

Garden of the sun, 108
Gnomon, 99, 130, 152, 153, 155, 156, 164, 209
God with scepter, 53, 54, 90
Golden scepter, 69, 70
Gregorian calendar, 131, 132, 155, 161, 165, 206, 207
Gregorian reform, 131

H

Hanan Cuzco, 71, 77, 78, 84, 95, 198
Heliacal rising of a star, 2, 47, 136, 139, 147, 149, 154-156, 173, 174, 177, 182-184, 206
Heliacal setting of a star, 47, 136, 139, 177, 182, 184, 207
Horizon astronomy, 151, 163
Huaqueros, 19
Huarochiri manuscript, 28, 169, 175, 178, 179, 221
Human sacrifices, 49, 50, 61, 91, 137, 140, 191, 201

I

Ice mummy (ice maiden), 20
Inca architecture, 93, 105
Inca comet, 190
Inca cosmogony, 67, 84, 198, 199, 202
Inca cosmovision, 27, 40, 87
Inca culture, 14, 15, 21, 192

Inca empire, 14, 17, 21, 23, 31, 33, 34, 58, 68, 69, 73–75, 77, 93, 94, 109, 110, 113, 115, 135, 137, 146, 159, 160, 184, 188, 191, 197, 199, 201–203
Inca pantheon, 77, 85–88, 144
Inca trail, 114, 116, 117
Inca world, 16, 156, 181
Inca zodiac, 183
Intercalation process, 133, 166
Intermediate periods, 1, 15, 34, 35, 40, 56, 70
Intiwatana, 103, 108, 118, 120–123, 147, 164, 199
Irrigation, 6, 40, 41, 48, 49, 57, 58, 71, 81, 86, 94, 116, 150, 161, 168
Islands of the sun and the moon, 111, 144, 146

J

Jesuits, 20, 24–28, 133, 178
Julian calendar, 131, 206, 207

K

Kamaq, 83, 147, 199
Kancha enclosure, 97
Khipukamayoq, (*see* master of Khipu)
Khipu, 20, 23, 28–31, 65, 79, 132, 133, 135, 155, 199, 201
King of Spain, 21, 23

L

La Niña, 11, 199
Lake Titicaca, 5, 8, 15, 23, 53, 54, 66, 70, 72, 77, 86, 87, 111, 121, 125, 126, 140, 144, 146, 159, 160, 191, 198, 200
Lanzón monolith, 36, 90

Subject Index

Legendary foundation of the Empire, 57, 70, 197
Local chief (kuraka), 1, 79, 193, 199, 200
Lord of Sipán, 19, 49, 50
Lord of Ucupe, 50

M

Machu Picchu site, v, 15, 16, 19, 102, 113–124, 146–150, 156, 160, 164, 172, 202
Major deities, 60
Masters of the Khipu (Khipukamayoq), 23, 28, 30, 31, 65, 79, 199
Mausoleum for mummies, 122
Messengers (chaskis), 30, 31, 198
Middle Horizon, 34, 56
Militarism of the Waris, 55
Milky Way, 2, 84, 86, 89, 124, 168–170, 172, 178, 180–184, 187, 194, 200, 201, 203, 207
Moche civilization, 35, 40, 49–51
Mochica theocracy, 49
Monastery of Santo Domingo, 109, 113
Moon, 2, 36, 46, 47, 60, 61, 77, 84–88, 106–108, 111, 112, 118, 129, 133, 144, 148–151, 155, 163, 166, 169, 183, 185, 187–189, 193, 199–201, 205, 207–209
Mummified heads, 40
Museo Arqueólogico Rafael Larco Herrera, 17, 18
Mythical creatures, 89, 90

N

Nadir (or anti-zenith), 131, 148, 153, 154, 161, 182, 208, 210
Names of the months, 133, 134

National Museum of Archaeology, Anthropology, and History of Peru, 16, 17, 19, 20, 74
Nature of wakas, 137
Nazca civilization, 34, 35, 40–46, 49, 192
Nazca geoglyphs, 42–44
Necropolis, 13, 14, 20, 39, 40, 62, 200

O

Orientation of monuments, 2, 143, 145, 146, 149, 151, 153, 154, 174
Orientation platform, 103

P

Pacific Ocean, 1, 5, 10, 11, 40
Paijan man, 16
Paracas culture, 34, 35, 39-42, 45, 62, 198, 200
Peruvian altiplano, 15, 54, 70, 72, 81, 88, 126, 127, 197, 198
Peruvian archaeologists, 13, 16–20, 36–38, 42, 44
Pilgrimage centers, 36, 106, 110, 144, 185
Pilgrimage of Coyllor Riti, 185, 201
Pillars (sukankas) near Cuzco, 155, 163, 164, 202
P'isaq site, 99, 100, 102, 103, 105-108, 122, 147, 148, 164
Pizzaro brothers, 98
Planet Venus, 88, 157, 178, 184, 185, 187, 194, 198
Plaza de Armas, 93, 96
Pleiades, 2, 61, 124, 136, 137, 139, 147, 149, 151, 154–156, 159, 160, 173–175, 177, 178, 184, 185, 187, 194, 198, 201
Pottery, 34, 39–41, 45, 46, 50, 58, 109

Pre-ceramic period, 15, 16, 18, 19, 34, 35
Pre-Columbian archaeology, 15
Pre-Hispanic Andean religiosity, 83
Princess's bath, The, 101
Puna, 5, 6, 10, 81, 86, 126, 191, 198, 201

Q

Qorikancha (Temple of the sun), 77, 78, 87, 88, 95, 96, 107, 108, 112, 114, 115, 135, 136, 138–141, 143, 149, 154, 155, 165, 178, 186, 201, 202
Quechua language, vii, 21, 23, 25, 26, 28, 62, 65, 71, 77, 94, 113, 133, 161, 168–170, 175, 176, 182, 183, 197, 201

R

Raimondi stele, 36, 90
Recent Horizon, 34, 35
Recent intermediate period, 34, 35, 70
Ritual of Khapaq Hucha, 137, 140, 141, 193, 201
Role of sukankas, 163, 164, 202

S

Sacred enclosure, 107, 201
Sacred place (waka), 16, 21, 22, 69, 83, 95, 106, 110, 133–140, 143, 145–148, 154, 161, 163, 178, 193, 194, 199, 202, 203
Sacred valley, 100, 147, 148
Sacrifice room, 99
Sacrificial knives (tumis), 56, 59, 202
Salinar culture, 17, 35, 38, 49
Sapa Inca, 72, 78, 79, 83, 87, 94, 104, 107, 135, 138, 140, 141, 146, 149, 163, 190, 193, 194, 197, 199–202
Secondary deities, 88
Selva, 1, 6, 202
Seq'es, 30, 78, 129, 133, 135–140, 143, 146, 147, 149, 154, 160, 163, 178, 198, 199, 201–203
Sican civilization, 35, 56
Sierra, 1, 6, 54, 90, 191, 192, 202
Site of Ollantaytambo, 104, 122
Sky observations, 2, 113, 161, 183
Society of Jesus, 24, 25
Solar and lunar eclipses, 61, 67, 87, 188, 189, 205, 207, 208
Solar towers, 165
Solstices, 2, 42, 44, 46, 47, 87, 97, 99, 101, 103, 105, 123–125, 131, 132, 134, 139, 140, 144–147, 149, 150, 152–156, 159, 161, 164–167, 169, 172, 182–184, 194, 199, 201, 202, 208, 209
Solstitial orientations, 99, 102
Sources of astronomy, 160
Southern cross, 123, 124, 139, 148, 168, 170–172, 177, 179–182, 187, 194, 214
Spanish chroniclers, 22, 60, 65, 66, 71, 78, 95, 100, 111, 114, 121, 132, 133, 161, 165, 169
Spanish conquest, 14, 15, 23, 65, 71, 74, 98, 135, 154, 161, 167, 198
Spanish priests, 24
Spondyls, 50, 90, 138, 140, 200
Spring equinox, 130, 131, 202
Stars, 2, 36, 46, 47, 61, 87–89, 103, 108, 122–124, 129, 130, 136, 137, 139, 143–145, 149, 151, 156, 159, 166, 168–179, 182, 184–187, 189, 194, 198, 200, 201, 206–209, 211
Summer solstice, 87, 122, 123, 130, 131, 145, 153, 163, 208

Sun, 2, 36, 42, 44–47, 55, 60, 61, 67–69, 78, 81, 83–87, 90, 95, 97–99, 103–108, 111, 112, 122–125, 129–132, 140, 143–146, 150, 152–155, 159, 163–167, 169, 177, 181, 183, 185, 187, 188, 194, 199, 205–209
Sun at the zenith, 132, 139, 147, 148, 152–155, 161, 209
Sun god, 70, 72, 86, 88, 96, 108, 118, 193, 199, 201
Sun observations, 2, 148, 150, 151, 194
Sunrise on the horizon, 152
Sunset on the horizon, 149, 152, 165
Synodic period, 129, 161, 166, 209
System of seq'es (or ceques), 78, 133

T

Tawantinsuyu empire, 66, 77, 93–95, 126, 202
Temple of the sun, 27, 99, 102, 107, 110, 118, 121, 122, 124, 133, 135, 136, 147, 149, 154, 163, 174, 178, 201–203
Temporal horizons, 1
Three universes, The, 144
Tiwanaku culture, 34, 35, 53, 55, 70, 85, 191, 192, 202
Torreón, 118, 121, 124, 147, 202
Towers of Chankillo, 145, 146, 221
Travelers and explorers, 13
Tropic of Cancer, 131, 153, 209
Tropic of Capricorn, 131, 153, 209
Tropical year, 3, 129, 130, 133, 139, 146, 152, 154, 163–166, 194, 200, 203, 207, 209
Types of waka, 21, 22, 69, 83, 95, 106, 110, 133, 135–140, 143, 145–148, 154, 155, 161, 163, 178, 193, 194, 199, 202, 203

U

Uranometria, 170
Urin Cuzco, 77, 95, 202
Uros people, 126, 128, 191
Urubamba river, 72, 100, 102, 104, 105, 114, 122–124, 147, 148

V

Viceroy, 21, 24, 68, 75, 202
Vilcanota river, 93, 100, 102, 105, 114, 168
Violent weather phenomena, 119
Virgins of the sun (akllas), 81, 197

W

Wari culture, 1, 34, 35, 43, 53, 55, 56, 61, 70, 85, 160, 191, 192
Watanay river, 71, 78, 93, 147
Wet and dry seasons, 8, 113, 124, 163, 169, 181
Winter solstice, 42, 44, 87, 98, 99, 123, 124, 130, 131, 145, 153, 166, 185, 208

Index of Proper Names, Rulers, and Deities

A

Acsumama, 88, 197
Ai Apaec, 49
Alaec pong, 61
Alcaviça, 69
Alcyone, 173
Allpa, *see* Mama Allpa
Alva, Walter, 19, 33, 34, 50
Amaru, 179, 181, 197
Amaru Inca Yupanki, 96, 197
Amaru Tupa Inca, 190, 197
Ancochincay, 177
Anello Oliva, Giovanni, 26
Anonymous chronicler, 28, 163, 167
Apu, 88, 136, 145, 197
Apu Inti, 87, 199
Apu Maya, 72
Aratus de Soles, 175
Arriaga, Pablo José, 26, 169, 174, 185

Asterope, 173
Atawallpa, 21, 71, 74, 75, 190, 197, 200, 203
Atlas, 160, 173
Avalos de Ayala, Luis, 23

Avendaño, Fernando (or de Avendaño, Fernando), 26, 169, 174, 175
Awqi Inti, 87, 199
Ayar Awqa, 68, 69, 197
Ayar Kachi, 68, 69, 197
Ayar Manqo, 68, 69, 197
Ayar Uchu, 68, 197

B

Bandelier, Adolph Francis Alphons, 15
Bayer, Johann, 170, 171
Berns, Augusto, 15
Bingham, Hiram, 15, 114
Bird, Junius, 16

C

Cabello de Balboa, Miguel, 24, 30, 57, 66, 177
Caesar, Julius, 131, 207
Carnegie, Andrew, 15
Cavillace, 88

Celaeno, 173
Ceterni, 57
Ch'aska (or Ch'aska Qoyllur), 87, 88, 170, 176, 178, 184, 198
Child Jesus, 10
Chimpu Ocllo, Isabel, 24
Choquechinchay, 177, 179, 187
Churi Inti, 87, 199
Cieza de León, Pedro, 22, 108, 154, 165, 190
Cium, 57
Cobo, Bernabé, 22, 24, 27, 30, 66, 78, 111, 132, 133, 135, 138, 139, 143, 147, 163, 166, 175–177, 179, 187, 189
Columbus, Christopher, 27
Coniraya, 88
Copacati, 88
Copernicus, Nicolas, 130

D

De Acosta, José, 24, 133, 177, 178
De Albornoz, Cristóbal, 24, 121, 133

De Arriaga Pablo, José, 174, 185
De Avendaño, Fernando, 26, 169, 174, 175
De Avila, Francisco, 28
De Houtman, Frederick, 170
De Lacaille, Nicolas-Louis, 171, 172
De la Calancha, Antonio, 27, 60, 165, 169
De la Gasca, Pedro, 22
De Matienzo, Juan, 22, 133
De Mendoza, Antonio, 21
De Molina, Cristóbal, 23, 66, 132–134, 139, 140, 167
De Montesinos, Fernando, 27, 165, 167
De Monzón, Luis, 28
De Murúa, Martin, 23, 30, 70, 133

De Santa Cruz Pachakuteq Yanki Salqamaywa, Juan, 27, 66, 184–186, 190
De Toledo, Francisco, 21, 24, 68
De Torres, Bernardo, 27
De Velasco y Pérez Petroche, Juan, 28
De Xerez, Francisco, 190
Diez de Betanzos, Juan, 21, 66, 132, 134, 135, 163, 165
Dirkszoon Keyser, Pieter, 170
Doña Angelina. *see* Ocllo Cuxirimay

E

Ekhako, 88
Electra, 173
Eqeqo, 88, 198
Eratosthenes, 153

F

Fur, 61, 174

G

Garcilaso de la Vega Inca (or Gómez Suárez de Figueroa), 24
Garcilaso de la Vega, Sebastián, 25, 26, 29, 70, 98, 104 , 132, 163, 165–167, 169, 184, 189
González Holguín, Diego, 26, 177, 185
Greer, Paolo, 15
Gregory XIII, Pope, 131, 132, 206
Guamán Poma de Ayala, Felipe, 23, 30, 67, 68, 70, 167, 169, 182, 184

H

Halley, Edmond, 190
Hesiod, 159, 160, 173
Hevelius, Johannes, 171, 176
Homer, 175

Index of Proper Names, Rulers, and Deities

I

Illapa, 86, 87, 199
Imaymana Wiraqocha, 66, 203
Inca Roq'a, 72, 96, 199
Inti, 70, 86–88, 96, 118, 193, 199
Izumi, Seichi, 18

J

Jiang, 61
Juanita ('the ice maiden'), 20

K

Kapa Yupanki, 72
Kauffmann Doig, Frederico, 20
Kepler, Johannes, 130
Killa, *see* Mama Killa
Kilya, 89
Kinwamama, 88, 199
Kon, 88, 199
Kroeber, Alfred Louis, 16
Kukamama, 88, 199
Kusi Quoyllur, 104
K'uychi, 88, 199

L

Larco Herrera, Rafael, 17, 18
Larco Hoyle, Rafael, 16–18, 37, 38
Lliwyaq, 86, 199
Lloq'e Yupanki Inca, 72, 96, 200

M

Machacuay, 177
Magellan, 24
Maia, 173
Mallku, 89
Mama Allpa, 89, 200
Mama Killa, 87, 88, 200, 201
Mama Nina, 89
Mama Oqllo, 68–70, 87, 126, 197, 199, 200, 202

Mama Pacha (or Pachamama), 84, 86, 201
Mama Qocha, 88, 200
Mama Qora, 68, 197
Mama Rawa, 68, 197
Mama Wayra, 89
Mangus, Chistophorus, 170
Manqo Inca, 75, 97, 98, 115, 200, 202
Manqo Qhapaq, 21, 68–70, 72, 87, 96, 106, 107, 126, 138, 144, 199, 200, 202
Mayta Qhapaq Inca, 72, 96, 138, 200, 201
Mejia Xesspe, Toribio, 140
Meneses de Alva, Susana, 19, 50
Merope, 173
Messier, Charles, 173
Minchançaman, 58
Muelle, Jorge C, 16, 17

N

Naymlap, 56, 57
Ni, 61
Nina, *see* Mama Nina

O

Ocllo Cuxirimay (or Doña Angelina), 21
Ollanta, 104
Oncoy, 174, 176
Oqllo (Mama), 68–70, 87, 126, 197, 199, 200, 202

P

Pachakamaq, 14, 55, 67, 68, 73, 84–86, 89, 109, 110, 200
Pachakuteq, 63, 67, 71–73, 83, 86, 87, 94, 96, 97, 101, 102, 104, 107, 109, 113, 115, 133, 135,

143, 144, 163, 165, 166, 197, 200, 202
Pachakuteq Inca Yupanki, 165, 190, 197, 200
Pachakuteq Yanki Salqamaywa, 27, 66, 169, —185, 186, 190
Pachamama (or Mama Pacha), 84, 86, 88, 101, 111, 201
Pariaqaqa, 88, 185, 201
Paricia, 89
Philip II, 21, 170
Piguerao, 89
Pizarro, Francisco, 74, 111, 190
Pizarro, Gonzalo, 21, 22
Pizarro, Hernando, 106
Pizarro, Juan, 96, 98
Pizarro, Pedro, 27, 98
Plancius, Petrus
    see van der Plancke, Petrus
Pleione, 173
Polo de Ondegardo y Zárate, Juan, 21, 22, 25, 27, 66, 132–134, 165, 166, 169, 174–177, 183, 187, 189
Poma de Ayala Felipe, Guamán, 23, 30, 66–68, 70, 134, 167, 169, 182, 184
Prescott, William H., 16
Ptolemy, 170, 171, 175, 177

Q

Qhapaq Yupanki Inca, 63, 72, 96, 201
Qocha, (see Mama Qocha) Qolca
Qon Teqsi Wiraqocha, 66, 68, 115
Qora (Mama), 68, 197
Qoyllur, 87–89, 168, 170, 176, 178, 179, 184, 198, 201

R

Raimondi, Antonio, 36, 90, 114
Rawa (Mama), 68, 197

Reiche, Maria, 44, 45, 47
Reinhardt, Johan, 19, 20, 118
Reiss, Wilhelm, 13
Rowe, John Howland, 14, 60, 61

S

Saramama, 88, 187
Sarmiento de Gamboa, Pedro, 24, 66, 68, 132, 133, 167
Sayri Thupaq Inca, 71, 75, 202
Shady Solis, Ruth, 19, 145
Shi, 60
Sinchi Roqá Inca, 69, 72, 96, 138
Sipán (Lord of Sipán), 19, 50
Stübel, Alphons, 13, 14
Suárez de Figueroa, Gómez, see Garcilaso de la Vega
Supay, 89, 202

T

Tambochacay, 69
Taycanamu, 58
Taygeta, 173
Tello Rojas, Julio César, 16, 17, 36, 39
Titu Kusi Yupanki Inca, 75
Tocapo Wiraqocha, 66, 203
Toledo, 75, 202
Tonupa, 105
Tulumanya (or Turumanyay), 89
Tupaq Amaru Inca, 21, 71, 89, 75, 97, 197, 202
Tupaq Inca Yupanki, 21, 58, 71, 73, 96, 97, 115, 138, 202, 203
Tutañamca, 89
Tycho Brahe, 170

U

Ucupe (Lord of Ucupe), 50
Uhle, Friedrich Maximilian, 14, 17, 110

Ukuku, 185
Urcaguary, 89
Urkuchillay, 88, 176, 177, 179
Urqu, 72

V

Valera, Blas, 25, 30, 185
Van der Plancke, Pierre (or Petrus Plancus), 171
Vázquez de Espinosa, Antonio, 26, 166
Virgil, 175
Von Humboldt, Alexander, 9

W

Waku, 68, 69, 197
Waskar Inca, 21, 71, 74, 75, 83, 96, 200, 203

Wayna Khapaq, 21, 24, 71, 73–75, 96, 97, 102, 104, 109, 138, 190, 200, 203
Wayra, *see* Mama Wayra
Wiener, Charles, 114
Wikatiraw, 72
Willaq Umu, 87, 203
Wiraqocha, 66–68, 71, 72, 85, 86, 96, 104, 105, 108, 118, 144, 203

Y

Yampallec, 57
Yanañamca, 89
Yawar Waqaq Inca, 72, 203

Z

Zamorano, Rodrigo, 170

GPSR Compliance

The European Union's (EU) General Product Safety Regulation (GPSR) is a set of rules that requires consumer products to be safe and our obligations to ensure this.

If you have any concerns about our products, you can contact us on

ProductSafety@springernature.com

In case Publisher is established outside the EU, the EU authorized representative is:

Springer Nature Customer Service Center GmbH
Europaplatz 3
69115 Heidelberg, Germany

www.ingramcontent.com/pod-product-compliance
Lightning Source LLC
LaVergne TN
LVHW010339260326
834688LV00036B/788